Making Liberia Safe

Transformation of the
National Security Sector

David C. Gompert, Olga Oliker, Brooke Stearns,
Keith Crane, K. Jack Riley

Prepared for the Office of the Secretary of Defense

 NATIONAL DEFENSE RESEARCH INSTITUTE

The research described in this report was prepared for the Office of the Secretary of Defense (OSD). The research was conducted in the RAND National Defense Research Institute, a federally funded research and development center sponsored by the OSD, the Joint Staff, the Unified Combatant Commands, the Department of the Navy, the Marine Corps, the defense agencies, and the defense Intelligence Community under Contract W74V8H-06-C-0002.

Library of Congress Cataloging-in-Publication Data is available for this publication.

ISBN 978-0-8330-4008-4

Cover Photo Courtesy REUTERS/Tim A Hetherington/Landov

Liberian police stand guard at an evening concert celebrating the inauguration of President Ellen Johnson-Sirleaf in Monrovia, January 16, 2006. Johnson-Sirleaf took office as Africa's first elected woman president, pledging to break with the country's history of corruption and violence that spread war to neighbouring states.

Published 2007 by the RAND Corporation
1776 Main Street, P.O. Box 2138, Santa Monica, CA 90407-2138
1200 South Hayes Street, Arlington, VA 22202-5050
4570 Fifth Avenue, Suite 600, Pittsburgh, PA 15213-2665
RAND URL: http://www.rand.org/
To order RAND documents or to obtain additional information, contact
Distribution Services: Telephone: (310) 451-7002;
Fax: (310) 451-6915; Email: order@rand.org

Preface

This report is the final component of the RAND Corporation's research project with the U.S. government under which RAND was asked to advise the Liberian and U.S. governments on security sector transformation in Liberia. This report should be of interest to the Liberian government, the U.S. government, the United Nations, other countries and organizations now engaged in reforming Liberia's security sector, and students and practitioners of security sector reform in general.

By agreement with the U.S. and Liberian governments, and by RAND's own tradition, the analysis and findings of this report are independent. Although RAND worked closely with both governments in performing this study, the results are not to be taken as the views of either government.

This research was conducted within the International Security and Defense Policy (ISDP) Center of the RAND National Defense Research Institute, a federally funded research and development center sponsored by the Office of the Secretary of Defense, the Joint Staff, the unified combatant commands, the Department of the Navy, the Marine Corps, the defense agencies, and the defense Intelligence Community.

For more information on RAND's International Security and Defense Policy Center, contact the director, James Dobbins. He can be reached by email at James_Dobbins@rand.org; by phone at 703-413-1100, extension 5134; or by mail at the RAND Corporation, 1200 South Hayes Street, Arlington, Virginia 22202-5050. More information about RAND is available at www.rand.org.

Contents

Figures

Tables

Summary

The security institutions, forces, and practices of the regime of Charles Taylor, Liberia's former president, met *none* of the essential criteria for a sound security sector: coherence, legitimacy, effectiveness, and affordability. They were meant to serve the regime, not the nation, and were controlled and used—rather, misused—by one man, mainly against Liberia's people and neighbors.

Yet even under new, able, and decent leadership, the old structures and ways are unworkable, wasteful, and confused, and they enjoy neither the trust nor the cooperation of the Liberian people at this critical juncture. It follows that Liberia must make a clean break, adopting a new security architecture, forces, management structure, and law.

The government of President Ellen Johnson Sirleaf has made security sector transformation a high priority, and the United Nations, the United States, and others are helping Liberia build new forces. What has been done and planned so far to transform the Liberian security apparatus is valid and important; this study raises no fundamental questions about the soundness of what is already under way.

At the same time, Liberia and its partners need an overall security architecture, accompanied by a strategy to create it. Without an architecture and strategy, setting priorities will become increasingly difficult; gaps, redundancies, confusion, and political squabbling over forces are likely. In offering an architecture and strategy, this study identifies additional measures, including additional capabilities, that would make Liberia's security sector more coherent, legitimate, effective, and affordable.

The starting point for this inquiry is an analysis of Liberia's security environment, which is complex, fluid, and fraught with risk. Liberia faces a present danger of growing lawlessness and poor public safety, owing primarily to its large pool of jobless and unschooled youth, whose only experience is fighting. If public safety and the rule of law are not established and maintained, odds are all too good that more severe domestic threats will arise. In particular, political opportunists, warlords, or criminal figures may lure and organize ex-fighters into armed groups beyond the reach of, and ultimately endangering, the state. Because this could happen quickly, capabilities that strengthen dissuasion and preempt internal threats are at a premium.

The risk of sudden threats from abroad cannot be excluded. Even if outright state aggression against Liberia is improbable in present conditions, the potential exists for incursions by insurgents operating from adjacent states and for use of Liberian territory by insurgents to attack those states.

In this environment, Liberia needs an *integrated security concept* to guide the formation and use of new forces and of new institutions to manage those forces. That concept should entail

- concentrating on known challenges of law enforcement and public safety
- dissuading, deterring, and—if need be—defeating any organized internal threats that may arise
- preparing to defend against external aggression by states or, more likely, by nonstate actors.

From this concept, we derive *core security functions*:

- regular policing
- protecting and developing transportation links, infrastructure, and natural resources
- protecting key officials
- preventing and responding to civil unrest
- preventing and defeating organized armed opposition, up to and including full-blown insurgency

- providing border and coastal security
- responding to outright aggression
- developing appropriate and mutually beneficial relationships with neighbors and other interested parties
- collecting intelligence to support these functions.

Liberian security forces, supported by intelligence capabilities, must be able to fulfill these core functions in a cost-effective manner. The Liberian National Police (LNP) and Armed Forces of Liberia (AFL) are Liberia's basic building blocks for performing these functions.

The primary missions of the LNP are (a) to prevent and fight crime and (b) to maintain public safety. These missions call for a light, but sizable, community-friendly police force that can earn the confidence and cooperation of the Liberian people. Anticipating occasional civil disorder, the LNP should also have a branch capable of riot control (e. g., the police support unit).

The primary missions of the AFL are (a) to safeguard the country against possible external threats and (b) to support internal security forces in defeating any insurgency or other internal threat for which Liberia's internal security forces prove inadequate on their own. At present, nonstate external and internal threats are more likely than threats from neighboring states. The size of the AFL is less important than that it be superior in quality and capability to foreseeable threats.

In view of Liberia's particular security demands, this basic force structure can be enhanced by including in the LNP a small mobile "swing" police unit capable of (a) helping regular police meet heightened internal dangers, (b) challenging armed groups that form in defiance of the state's authority, and (c) operating with the AFL, if need be, to meet major internal or external threats. This quick-response police unit (QRPU) should be oriented toward law enforcement but also trained and equipped for light combat operations. QRPU personnel would be drawn mainly from the rest of the LNP; rotation of personnel through the LNP, including tours in the QRPU, would facilitate interoperability.

This overall architecture should provide flexibility, speed, and geographic reach. The QRPU would permit the regular LNP to be lightly armed and community-oriented, and it would reduce the state's reliance on AFL intervention to quell domestic threats.

Liberia's core security functions also indicate a need for a modest Coast Guard, in addition to the Customs and Immigration services and the Special Security Service (SSS) to protect national leaders.

In analyzing specific *force-structure options* (detailed in Chapter Four of this volume), we found the following (see Table S.1):

- Existing plans of the United Nations (UN), the United States, and Liberia to build a small LNP and small AFL (Option 1), while sound, may not be adequate to meet Liberia's needs—especially for maintaining basic public safety, preventing armed internal opposition, and providing coastal security.
- Doubling the planned size of the LNP and the AFL and adding a Coast Guard (Option 2), which would result in approximately $18 million more in annual operating costs, could fall short of providing adequate security against armed internal opposition without excessive reliance on domestic intervention by the AFL.
- Incorporating a QRPU in the LNP (Option 3) would better meet Liberia's security challenges, especially armed internal opposition, at a $5 million increase in annual operating costs above the current plan.
- Although the capital cost of Option 3 would be about $24 million more than that of Option 1—because of the addition of a QRPU and Coast Guard—it would cost $43 million less to build than Option 2. This seems like a wise investment for Liberia and its supporters, yielding effective security on an economical operating basis.

Table S.1
Costs for Three Force-Structure
Options ($millions)

Option	Operating Cost	Capital Cost
1	17.8	94.9
2	35.4	162.1
3	22.5	118.9

The force structure of Option 3 covers the full spectrum of internal and external dangers, including those from armed gangs and insurgency. At the same time, the ability of Liberian security forces to meet these dangers can be affected by poor road access, aggravated by the difficulty in moving during the rainy season. This problem can be reduced by good surveillance, rotary-air mobility (provided by a foreign partner), preemptive action, and isolation of armed groups in inaccessible areas, as well as by the ability to act in force when roads become passable. Improving Liberia's roads is important for its security.

The United Nations Mission in Liberia (UNMIL) is critical to Liberia's security and will remain so for some years to come. It will take about five years before the main Liberian forces have been fully built, equipped, trained, and deployed.[1] During that period it should be possible to scale back significantly the numbers of UNMIL peacekeepers and correspondingly reduce UNMIL costs, provided certain critical UNMIL capabilities are preserved—especially police advisors, UNMIL's own quick-response force, and rotary-wing air transport and surveillance. During this transition, command and control arrangements between UNMIL and Liberian security forces must be delineated and coordinated with great care.

Although it is unclear whether the UN will be prepared to maintain any presence beyond the time Liberian forces reach full strength, a tailored residual presence, on the order of no more than 3,000 personnel, could be needed for at least one or two years thereafter, given stable conditions, to ensure that conflict and chaos do not return. Beyond that time and for some time to come, a small but critical need will remain for international (not necessarily UN) capabilities, including advisors, rotary-wing air transport and surveillance, to complement Liberian forces. The cost of such a post-UNMIL international capability has been estimated in the body of this monograph; it is not included in the Liberian force-structure options.

Because Liberia's security environment is dynamic and unpredictable, force plans and the force structure itself must be adaptable.

---

[1] This assumes a benign environment and significant continued and new assistance. The time frame is therefore somewhat notional.

This goes not only for the mix of capabilities of Liberian security forces—e.g., the size of the regular police, the relative importance of the QRPU, the size and firepower of the army—but also for the rate at which UNMIL can be drawn down. This demands tight planning links between the Liberian government, the U.S. government, and the UN. Liberia must develop its own ability to plan its needs for forces and other security capabilities based upon informed, objective, and realistic analysis. It must neither underestimate the security difficulties it faces nor overestimate its ability to maintain capabilities. Creating a civilian and military capability to assess, plan, and align its resources with its needs should become part of the assistance Liberia receives from its international partners in the coming years.

As important as Liberia's forces are its security institutions—the management structures, responsibilities, authorities, processes, and rules—that will assure coherent, legitimate, effective, and affordable direction to, control of, and support for security forces. These institutions are needed not only for Liberia's long-term security but also to guide security sector transformation starting now. The following merit immediate consideration:

- A Liberian National Security Council (NSC) for policymaking, resource allocation, and crisis management should be created and used regularly. The NSC would be chaired by the President and would include as its core the Ministers of Justice, Defense, Finance, and Foreign Affairs (with others included ad hoc). It would receive objective analysis and advice from the head of national intelligence, the most senior officers of the LNP and AFL, and the Liberian National Security Advisor (LNSA).
- This cabinet-level NSC should in turn serve as a template for, and should insist upon, interministerial information-sharing and coordination at lower levels—a bureaucratic challenge for any government, but essential for Liberia. Multilevel interministerial cooperation will take time to effect; all the more reason to encourage it now.
- The LNSA should have several responsibilities: orchestrating the NSC system at and below the cabinet level; ensuring that the Pres-

ident and NSC receive objective analysis, options, and all points of view; fostering direct ties among key ministries and agencies; monitoring the progress of security sector transformation; and monitoring the quality of operational cooperation among the various security services. The LNSA should not be involved in regular ministry affairs or come between ministers and the President.

- The chain of command over the AFL—the country's strongest force—should be clarified: from the President, as commander-in-chief, through the Minister of Defense to the senior military commander, with the understanding that decisions to use military force should be reached by the deliberation of the NSC as a whole. Any military domestic intervention, moreover, would require consultation with the legislature.
- Several ancillary police should be consolidated into the LNP, with the exception of certain specialized services—Special Security Service for executive protection, Immigration and Naturalization, Customs, and the Coast Guard.
- Other ancillary police agencies should be eliminated and their personnel vetted for possible service in the LNP.
- The LNP should be aligned under the Justice Ministry while maintaining operational control within the LNP, with an independent board to maintain professional standards and public trust of the police.

Intelligence capabilities are an essential complement of the various armed services and must be held to the same standards of effectiveness, affordability, legitimacy, and coherence. Responsibility and capability to collect intelligence should be concentrated in a National Security Agency (NSA) that (a) reports to the President; (b) provides analysis to the entire NSC; (c) furnishes information directly and continuously to the LNP and AFL; and (d) is authorized to arrest and briefly detain only persons who pose national security threats. Thus, the intelligence service is a *support* organization for the rest of the security sector. Recognizing that the police will be able to collect much of the information needed to investigate and fight crime, the NSA should focus on high-threat concerns and can be of modest size.

Taking this analysis of force structure and institutions into account, it is possible to assemble a complete architecture, as shown below. This architecture would have the following characteristics (see Figure S.1):

- The NSC, chaired by the President as commander-in-chief, has final authority over all security forces.
- Security forces report through ministries rather than directly to the President.
- Security forces are distributed between the Justice and Defense ministries.
- Lines of authority are clear.
- Control over the military passes from the President through the Minister of Defense.

Figure S.1
Integrated Architecture and Core Functions

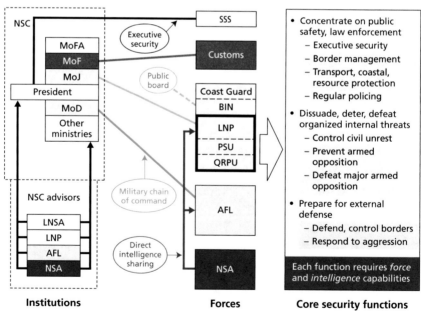

RAND MG529-S.1

- The number of distinct security forces and services is manageably small, while still allowing for specialization.
- No security force lacks an important core security function.
- No core security function lacks a force that is principally responsible for it, and there is no confusion or duplication in the alignment of forces with functions.
- The QRPU can support other police units or support the AFL.
- The intelligence service (NSA) reports to the President, serves the NSC as a whole, and provides direct support to the LNP and AFL.

This architecture should be presented and codified in a way that secures broad political support, earns public understanding and trust, and signals the government's clarity of purpose and resolve, including toward potential adversaries. A new omnibus national security law, though a political challenge to enact, is a better way to meet these needs than revising each law now on the books or instituting a new system by presidential decree.

In the course of preparing this integrated approach, we identified several issues in need of immediate attention:

- Security personnel should stay out of politics, except for having the right to vote.
- To avoid discontinuity and political manipulation, senior sub-cabinet security officials (except for the LNSA) and uniformed officers should be nominated by the President and confirmed by the legislature for fixed terms, not changed with a government transition.
- As the new LNP is being built, its patrols need to be accompanied and mentored by armed international civilian police (CIVPOL) advisors.
- Liberian justice, courts, and corrections systems must be built quickly or law enforcement will be neither effective nor legitimate; this effort is now woefully under-resourced.

- Personnel of the former police force and other security forces who are not to be trained and integrated into the new force should be retired immediately, lest they infect new police with bad habits.
- Current systems for paying security personnel must be upgraded and made immune to corruption.

Liberia must not and need not be left to face its dangers alone. Even as Liberian forces gradually take over from UNMIL, and as new security institutions are built, those with a stake in Liberia's security—the UN, the AU, the Economic Community of West African States (ECOWAS), the United States, other countries, and international organizations—should continue to help. Liberia should forge cooperative ties with its neighbors in the Mano River Basin, including coordination against common nonstate threats. The UN Security Council should make clear through continuing resolutions that its concern for Liberia will not fade with the gradual reduction of UNMIL. The United States must be steadfast in its support for Liberia, making it a model of how a failed state can be made secure and viable. As others offer to help Liberian security sector development, their efforts should conform to Liberia's chosen principles, architecture, and standards.

Implementation should focus on the following:

- Immediate and consistent use of the NSC to guide security sector policy, planning, resource-allocation, and transformation
- Development and coordination of detailed integrated (UNMIL-Liberian) force plans with the United States and the UN
- Public education, political consensus-building, and preparation of a national security law
- Stepped-up and regular joint LNP-CIVPOL patrolling to solidify the rule of law, provide evidence of deterrence, and show that the state is making progress
- Plans to assure uninterrupted continuation of UNMIL's own quick-response force
- A design and plans for a small LNP QRPU and small Coast Guard

- Consolidation, reduction, and appropriate recruiting, vetting, and training of the currently independent ancillary police forces, customs, and intelligence personnel
- Attention to building court and corrections-system capacity
- Training (e.g., at the U.S. Africa Center for Strategic Studies and other institutions) of senior officials and officers in the precepts and practicalities of Liberia's security sector.

To conclude, the presence of UNMIL, the commitment of the United States, and a somewhat less hostile external security environment—albeit one that may change rapidly—provide Liberia with valuable time to create security institutions and forces that are effective against dangers, are legitimate in the eyes of Liberia's people and neighbors, fit together and work coherently, and are worth the cost. This study is meant to help Liberia and its supporters use that time well.

Acknowledgments

Numerous individuals made significant contributions throughout the course of this project. Above all, we want to express our admiration and encouragement for those Liberians, starting with President Ellen Johnson Sirleaf, who are working tirelessly to bring peace to their country and their region. The President and her team of ministers and advisors were generous with their time and their thoughts, for which we thank them first of all. We are also particularly grateful to Theresa Whelan and Vic Nelson, who encouraged and supported the project from the outset. At the State Department, Linda Thomas-Greenfield, Peter Davis, Allison Henry-Plotts, and Susan McCarty provided substantial assistance. John-Peter Pham provided invaluable consultation in the beginning stages and offered frank and insightful comments throughout the project, including a review of the draft final report. We are also grateful to Michael McGovern, Karin von Hippel, James Dobbins, LtCol Mark "Duke" Ellington, Andy Michels, and Dmitri Titov and his staff at the United Nations, for their important insights.

This report was informed by field visits to Liberia and exchanges with hundreds of stakeholders and experts both in Liberia and elsewhere. We are grateful to members of the Liberian government, civil servants, UN officials, U.S. Embassy staff, international and Liberian nongovernmental organizations, and civil society members who informed our work. In particular, we would like to thank Donald Booth, Alfreda Meyers, Tony Yowell, and Dan Honken of the U.S. Mission; Alan Doss, the Special Representative of the UN Secretary General (SRSG); and Allison Kemp. Rob Deere of the SRSG's staff provided extensive assistance in facilitating trips to outlying areas in Liberia as well as to Sierra Leone, and in providing UNMIL's perspectives—through discussions and a formal document—on security sector reform.

At RAND, John Gordon provided invaluable technical assistance, while Adam Grissom offered a candid and insightful review, which greatly improved the report. Lesley Warner, Nathan Chandler, and Sarah Harting all contributed excellent administrative support.

Any errors of fact and interpretation in this report are solely the responsibility of the authors.

Abbreviations

AFL	Armed Forces of Liberia
ATU	Anti-Terrorist Unit
AU	African Union
BIN	Bureau of Immigration and Naturalization
CIVPOL	international civilian police
CPA	Comprehensive Peace Agreement
ECOMOG	Economic Community's Monitoring Group
ECOWAS	Economic Community of West African States
FDA	Forest Development Authority
INPFL	Independent National Patriotic Front of Liberia
LNP	Liberian National Police
LNSA	Liberian National Security Advisor
LPRC	Liberia Petroleum Refining Company Security Force
LSP	Liberian Seaport Police
LURD	Liberians United for Reconciliation and Democracy
MCP	Monrovia City Police
MNS	Ministry of National Security
MoD	Ministry of Defense
MODEL	Movement for Democracy in Liberia
MoF	Ministry of Finance
MoJ	Ministry of Justice
NBI	National Bureau of Investigation

NPFL	National Patriotic Front of Liberia
NSA	National Security Agency
NSC	National Security Council
PSU	Police Support Unit
QRPU	Quick-Response Police Unit of the LNP
RIA	Roberts International Airport Security Force
SRSG	Special Representative of the UN Secretary General
SSS	Special Security Service
UNMIL	United Nations Mission Liberia
UNPOL	UN police

CHAPTER ONE

Introduction

After fourteen years of conflict and two years of transitional rule, Liberia has a democratically elected government committed to creating a peaceful future for the country and its people. Through decades of dictatorship and civil war, the government and its security forces came to be seen by most Liberians as perpetrators of violence, masters of corruption, and abusers of power. The new Liberian state must jettison past security practices, institutions, and forces, not only to provide for its people's safety but also to gain their trust and cooperation. Simply reforming its security sector is not enough; it must build a new one on the political, moral, and physical rubble of the old.

Making a clean break is crucial, but that alone is not enough. Liberia's viability as a country depends on its being at peace with itself and with its neighbors, which in turn depends on the details of *how* and *how well* the state provides for public and national security. The institutions, forces, and practices of Liberia's transformed security sector must be coherent, legitimate, effective, and affordable.

Liberia has important friends as it strives for a peaceful future and a sound security sector. The United Nations has made major contributions, including demobilizing Liberian combatants, deploying a large peacekeeping force (UNMIL), and training police. The United States has said that it will do all it can to help Liberia recover and is providing considerable aid, including building a new Liberian army. Other countries and international institutions are lending a hand out of their conviction that a democratic Liberia deserves support and is important for the stability and development of its neighbors and West Africa as a whole.

Although having numerous good partners is invaluable for Liberia, they compound the difficulty of formulating a coherent and agreed-upon strategy for rebuilding, especially in the national security arena. Conversely, with the combination of a coherent strategy, Liberian–U.S.–UN teamwork, sustained and well-focused international support, leadership, and broad national will, the odds for success are good.

UNMIL is currently providing security for Liberia. Yet the government of Liberia, wisely, is not deferring the creation of institutions and capabilities that will enable the Liberians to provide for their own security in the future, even if international support continues. Drawing down and eventually withdrawing UNMIL depends on the success of Liberia and its partners, especially the UN and the United States, in building Liberian institutions and capabilities.

Concrete progress has already been made, notably,

- demobilization and disarmament of some 100,000 Liberian fighters
- training, thus far, of approximately 2,000 police by UNMIL
- the beginning of recruitment for a new army with U.S. help
- the appointment of capable ministers, officials, and officers to lead new security institutions and services.

These efforts are important, and this study raises no basic questions about their validity. Missing, however, is an integrated analysis of the security sector institutions, forces, and services needed for the new Liberia—an overall architecture in which they can fit and a unified strategy for creating them. Such a "big picture" is important to maximize the benefits from individual investments and efforts and to set priorities for the use of scarce resources. RAND was asked by the U.S. and Liberian governments to furnish this missing analysis.

To that end, this report identifies and analyzes options and issues covering all key aspects of Liberian security sector reform, from the national policy and decisionmaking apparatus to the forces and services that will maintain security, as well as the links of authority, guidance, and accountability that make these forces and services responsive to the elected government and to the people.

Subsequent to this introduction, the report is structured as follows:

- Chapter Two reviews the recent conflicts that have engulfed Liberia, insofar as its violent history casts a forward shadow on efforts to build a new system. It then analyzes Liberia's current and likely future security environment; in that light, it suggests a national security concept and several core security functions.
- Chapter Three presents assessment criteria and enduring principles to be used to design an overall architecture and to evaluate options for the various institutions and services.
- Chapter Four lays out several integrated force-structure options, encompassing police, military forces, and other armed elements, and analyzes and compares the cost-effectiveness of these options. It then links the most cost-effective of these force options with long-term plans for UNMIL and other international security support. This chapter also provides estimated investment and operating costs for the options, as well as approximate costs of a continuing international presence.
- Chapter Five begins with a diagnosis of the current governmental organization of Liberia's security sector and provides options for improved decisionmaking, clear command authority, and options for organizing police, borders, coast guard, and intelligence structures.
- Chapter Six identifies and analyzes other issues for consideration, including policing priorities, justice and courts, personnel, and international security cooperation.
- Chapter Seven concludes the monograph by summarizing key findings and highlighting implementation steps that demand prompt attention from the Liberian government.

As a working definition of "security sector," we include those institutions, forces, and other services, decisionmaking structures, laws, and policies that provide, operate, and resource the capability of the state to protect the nation's people, resources, territory, and elected government from internal and external dangers. These include, among others, police and other law enforcement services; military forces; intelligence functions; the ministries of Justice, Foreign Affairs, Finance, Defense, and possibly others; cabinet-level policymaking; and interministerial cooperation.

The justice, court, and penal systems are critical to the security sector. The mechanisms for allocating and accounting for resources and general government administration at the national and subnational levels are also key factors in creating an effective security sector. The security sector also depends on the cooperation of the Liberian people, which must be earned over time by effective and trustworthy security forces and policies.

Throughout this document, the terms "security sector reform," "security sector transformation," and "security sector development" are used interchangeably. Whatever the terminology, this study reveals that Liberia's security sector is in need of sweeping change.

CHAPTER TWO

Security Environment and Demands

Background

For most of its history, Liberia was a relatively peaceful one-party state, with a multiethnic indigenous majority ruled primarily by the America-Liberian minority. Unlike many of its neighbors, Liberia did not have a martial tradition. It was not until the beginning of the twentieth century that Liberia created the Liberian Frontier Force, primarily to prevent British and French colonial powers from encroaching on its borders.[1]

In the 1960s, with support from the U.S. government, Liberia developed the Armed Forces of Liberia (AFL). This force quickly grew to a 6,000-strong national military with a National Guard and Coast Guard. The officer corps of the AFL consisted almost exclusively of America-Liberians, and indigenous Liberians served as soldiers and noncommissioned officers. When President William Tubman died in 1971, then-Vice President William Tolbert became President of Liberia. Although he succeeded in breaking down the America-Liberian patronage system, his reforms did not benefit most Liberians quickly enough. Not only did Tolbert lack support from the general populace, he also earned dissatisfaction among the AFL. Tolbert alienated the AFL by removing officers on charges of disloyalty.

---

[1]   Pham (2004).

In the context of growing unrest and poverty, strong opposition to Tolbert was heightened when he substantially raised the price of rice, Liberia's staple food. Riots ensued, and the government ordered security forces to respond. Not only did the situation end with a massacre, but Tolbert offended many soldiers by utilizing Guinean troops to respond to this domestic situation.[2] This incident was the beginning of the "Year of Ferment," marked by oppression of political opposition and public demonstrations.

In the aftermath of this period, there was a nationwide strike demanding Tolbert's resignation. In 1980, Master-Sergeant Samuel Doe murdered Tolbert and seized power. Doe, a young Krahn (one of Liberia's ethnic groups) from Grand Gedeh County with little education, led a coup d'état with sixteen junior officers and privates. Doe's military regime was incompetent, repressive, corrupt, and ethnically divisive. His system of patronage, which primarily benefited his fellow Krahn, led to large government pay raises, significant increases in the number of Liberians on government payrolls, and rising international debt. By the mid-1980s, Doe was facing rising opposition, including a 1985 invasion from Sierra Leone led by General Quiwonkpa, Doe's former compatriot in the 1980 coup whom he had promoted to be commanding officer of the AFL.

Recognizing the lack of qualification and capacity among government employees, Doe had brought young technocrats into the government—who proved chiefly capable of gaining materially from their new positions. One such technocrat, Charles Taylor, was in charge of procurement, but fled Liberia in 1983 to avoid prosecution on charges of embezzlement of state resources. After fleeing Liberia, Taylor was found in the United States and was arrested and detained on the authority of a U.S.-Liberian extradition treaty. He broke out of Massachusetts' Plymouth House of Corrections and made his way back to Liberia.[3]

In December 1989, Taylor and his National Patriotic Front of Liberia (NPFL) invaded Liberia. With backing from Côte d'Ivoire and

---

[2] Sesay (1999), pp. 145–161.

[3] Pham (2004).

Burkina Faso, Taylor and his group of less than 200 men—recruited from unemployed, poorly educated, and dissatisfied Liberians in Nimba County, Northern Liberia—started a civil war that eventually resulted in approximately 200,000 deaths and the displacement of one million Liberians.[4]

With support from Guinea, Nigeria, and Sierra Leone, the Economic Community of West African States (ECOWAS) established the Economic Community's Monitoring Group (ECOMOG) as a peacekeeping mission in Liberia. Although ECOMOG initially succeeded in preventing Charles Taylor from seizing Monrovia in 1990, its long-term effectiveness was limited. Taylor expanded the war into Sierra Leone territory and took over the Monrovia suburb of Paynesville. When Taylor's NPFL cut off the water and electrical supplies to the capital, Guinea and Sierra Leone responded by supporting Doe. The NPFL began to divide, and Prince Johnson broke off and formed his own rebel movement, the Independent National Patriotic Front of Liberia (INPFL). Despite ECOMOG's presence, Prince Johnson and INPFL captured and killed Doe in 1990.

A battle to seize Monrovia ensued, and large number of refugees fled the country. New rebel groups, such as the United Liberian Movement for Democracy, formed and joined in the conflict. The United Liberian Movement for Democracy and ECOMOG succeeded in reducing the amount of territory under Taylor's control. Taylor responded by shifting from conventional assaults to surprise attacks on ECOMOG.

In 1997, after several years of strife, Taylor was elected President of Liberia in a contest that was marred by some irregularities and conducted in a climate not entirely free from fear of renewed violence. His regime's unprecedented levels of brutality, corruption, incompetence, intrigue, and foreign adventures completely broke the already fragile Liberian spirit and economy. Taylor's privatization of the state's rural resources (such as the timber, diamond, and mining industries), produced large gains for a select few and no investment in the state or

---

[4] ICG (2004).

social services for the Liberian people.[5] To maintain control over the country and the various security forces, Taylor replaced Krahn AFL members with troops loyal to him and created new security forces reporting directly to him.

Armed rebel groups, particularly Liberians United for Reconciliation and Democracy (LURD) and the Movement for Democracy in Liberia (MODEL) opposed Taylor in a civil war that lasted several years. By the end of May 2003, LURD and MODEL had gained control of more than 60 percent of Liberia's national territory.[6] Under mounting international, and particularly U.S., pressure, Charles Taylor resigned in August 2003, and a Comprehensive Peace Agreement (CPA) was signed. The CPA installed a national transitional government, chaired by Gyude Bryant, and the United Nations Security Council authorized a UN Mission in Liberia (UNMIL) that is currently 15,000 people strong.

In November 2005, Ellen Johnson Sirleaf was elected President of Liberia in a democratically free and fair election. She took office in January 2006. The cabinet has been appointed and approved, and initial steps have been taken to rebuild the failed state of Liberia.

General Environment

In light of its own violent background, recent conflicts in Sierra Leone and Côte d'Ivoire, and the potential for turmoil in Guinea, there is no question that Liberia is situated in a volatile part of Africa. Weak governance, ethnic feuding, competition for resources, nonstate armies, and cross-border meddling and incursions have undermined security in Liberia and its neighbors.

Liberia itself has been a source of much of this strife and conflict. The previous Liberian regime inflicted violence both within and across Liberia's borders, destroying both internal and external security and ultimately the state itself. The new Liberia must be the opposite of

---

[5] Sesay (1999).

[6] Pham (2004).

Charles Taylor's Liberia: at peace with itself, at peace with its neighbors, and ultimately a pillar of stability. With Taylor in custody awaiting trial for war crimes, and with Liberia and Sierra Leone, at least, on the path to recovery, the immediate region, though still fraught with potential danger, has an opportunity for improved security.

Liberia's internal security environment also has been improved by the presence of UNMIL and the demobilization of Liberian fighting forces. This internal environment contains more certain and immediate dangers than does the external one because of the serious potential for ex-combatants to be organized by one or another faction or warlord in opposition to the new state. Taylor's legacy is a country still perilously close to the abyss of lawlessness—lawlessness that domestic enemies of democratic Liberia could be quick to exploit.

Liberia's internal security environment is certain to grow restive, violent, and chaotic if effective Liberian institutions and forces are not built and in operation by the time UNMIL begins to withdraw, if not sooner. In turn, to the extent that internal security is not maintained, turmoil in Liberia is almost sure to aggravate regional security conditions, either by spilling over into Liberia's neighbors or by causing foreign actors to see opportunity in renewed Liberian violence. The absence of external threats could change—and change rapidly—if Liberia cannot maintain domestic peace or if conditions turn worse beyond its borders.

With this complex and fluid security environment in mind, Liberia must give its immediate attention to confronting known internal dangers while also being prepared to face external ones.

Internal Security Challenges

Liberia faces two significant internal dangers: (1) widespread lawlessness and (2) the risk that alienated and dissatisfied Liberians will form into militias, rebel forces, or other armed groups. The potential for widespread and increasingly violent crime—theft, looting, battery, disorderly conduct, and killing—is just beneath the surface of Liberian life. A lack of electricity facilitates criminal activity during the night, especially because police officers are not adequately equipped with flashlights and bicycles for evening patrols. In addition, unarmed and still under-trained police officers have difficulty confronting criminals who are armed with cutlasses and other lethal weapons.

Lawlessness undermines public confidence in democratic government. It can lead to an erosion of general respect for law and to larger internal security problems, insofar as armed groups believe they can exploit a void with little risk. Failure to provide basic public safety can spawn new militias to protect one interest or another, even if they are not in direct opposition to the state. In addition, lawlessness deters investment, which is vital for the country's economic growth and ultimately for its stability and security. In sum, primary law enforcement must be a major and immediate concern of security sector reform— one that the Liberian state itself must provide and must be seen to be providing successfully.

Making law enforcement effective will, of course, be a real challenge even for a capable, professional, well-led police force. Liberia has large numbers of young, uneducated, and unemployed males who have known little but conflict. The U.S. Department of Defense found that 40 percent of the ex-combatant population consisted of males under the age of 16 who had never attended school.[7] Large numbers of demobilized ex-combatants are currently waiting to see what happens under the new Liberian government.

Liberian politics are not entirely settled. Some figures who enjoyed power in the past may feel the process is unfavorable to them. Antidemocratic and factional figures may regard the new order as a new opportunity for seeking power and lucre. Liberia has natural resources that can be exploited by lawless individuals, and ex-combatants are already taking advantage of these opportunities. For example, ex-LURD fighters have seized the Guthrie rubber plantation and taken over its operation, creating a virtual fiefdom with tolls and demands on the local populations. Disaffected youth, ex-combatants, and other Liberians with limited prospects for the future could be recruited to form organized armed groups with political or economic aims. Although large-

---

[7]   MPRI (2004).

scale armed opposition is unlikely with UNMIL present, rebel group structures and command chains have not been eradicated and remain a concern.

These two primary internal security threats create several derivative concerns for the Liberian government. First, the safety of Liberia's elected leaders and key officials cannot be taken for granted. Second, major natural resources and routes have to be secured as well as possible. Third, the security of the general population must be improved and maintained; the state's security apparatus must not be seen as dedicated only to protecting national leaders and economic interests, as important as they are.

Internal security is best provided through the unified efforts of the state and the people. Lack of public confidence in the security forces or a perception that the forces exist to serve only the state will damage the faith of the people in the new government's agenda—and the people will then withhold their cooperation in achieving security. The history of corruption among Liberian police officers is a prime example. Because officers have been corrupt, the public distrusts the police, making the problem of law enforcement much harder. In a democracy, the law cannot be enforced through coercion and fear. The provision of internal security must be as legitimate as it is effective. The more legitimate it is, the more effective it will be; the more effective it is, the more legitimate it will be.

External Security Challenge

Liberia's internal and external security, like that of its neighbors, has been intertwined, as armed groups have formed and trained in one country from where they attack another.

Currently, the Mano River Basin, which consists of Guinea, Liberia, and Sierra Leone, is somewhat more quiescent, although still volatile. Although no state is at present hostile toward Liberia, each of its immediate neighbors is potentially unstable.

- Sierra Leone appears to be leaving behind its history of conflict, although some observers say that the problems that led to the conflict have yet to be resolved.
- Guinea is currently stable; however, political uncertainty will increase as the health of the current president deteriorates. The ICG anticipates a political transition prior to the scheduled 2007 legislative elections.[8]
- In Côte d'Ivoire, there are still flare-ups of violence; a return to general conflict is very possible.
- The large numbers of fighters of one sort or another in both Côte d'Ivoire and Guinea pose a serious potential threat to Liberia, especially in the event of a regime change or collapse in those countries.

One reason for cautious hope for a less violent regional environment is that Charles Taylor's Liberia was a major source of insecurity, unrest, and violence in the region. A democratic, effective, and responsible Liberia removes a major cause of regional instability. At the state level, Liberia's neighbors should have no legitimate reason either to fear or to menace. If antagonism and conflict do arise, it is more likely to be because of difficulty in controlling border regions, some of which are rich in resources, against nonstate enemies.

However, as has often been the case in the past, Liberian territory could be used by rebel groups, potentially including Taylor loyalists preparing to attack neighboring countries. Similarly, actors plotting against the Liberian state could easily operate in neighboring countries. Liberia has long, porous borders with Sierra Leone, Guinea, and Côte d'Ivoire. It is infeasible for Liberia to completely control its borders, territory, littoral, and air space. The flow of people and goods across its borders and from the sea pose a serious security challenge; Liberia's borders must be monitored. Border and coastal security are critically important for both Liberia's security and its economic development and growth.

---

[8]   ICG (2006b).

Security Concept and Core Functions

In such a complex and unstable environment, Liberia needs an integrated security concept to guide the formation of new forces and new institutions to manage those forces. The concept should entail the following three elements:

1. Concentrating on the known challenges of law enforcement and public safety
2. Dissuading, deterring, and, if need be, defeating any organized internal threats that may arise
3. Preparing to defend against external aggression by states or, more likely, by nonstate actors.

Each element affects the others. Success in the first will make achieving the second easier. In turn, stronger internal security will make external threats less likely. Preparing to defeat outside threats will limit the potential for internal dangers to exceed Liberia's security capabilities.

From this concept, we can derive the following core security functions:

- Regular policing
- Protecting and developing transportation links, infrastructure, and natural resources
- Protecting key officials
- Preventing and responding to civil unrest
- Preventing and defeating armed opposition, up to and including full-blown insurgency
- Providing border and coastal security
- Responding to outright aggression
- Developing appropriate and mutually beneficial relationships with neighbors and other interested parties
- Collecting intelligence to support these functions.

The concentration on public safety and law enforcement will require regular policing by effective forces that have earned the trust and cooperation of the people. The police will be by far the most visible aspect of the state's commitment to providing security in an effective and legitimate way. Although it is important for Liberia's own police to take over law enforcement from UNMIL soon, the importance of good training suggests that it should not be rushed.

Effective regular policing, along with adequate protection of the country's economic security and leadership, will in turn benefit the second element of the security concept: the prevention of and response to organized internal threats. Such threats are far more likely to arise in a lawless environment, with its opportunities for assassination or seizure of resources. Conversely, the greater the faith of the Liberian people in the capabilities and behavior of their own police, the less they will be attracted to join or support armed gangs and opposition forces. The people are the best ally of any democratic state, especially in combating internal enemies.

The disarmament, demobilization, and rehabilitation process is incomplete. Although 30,000 ex-combatants were expected to participate, more than 100,000 individuals were disarmed. This unexpectedly large disarmament left very limited financial resources for rehabilitation. It is commonly believed that ex-combatants and rebel groups have weapons caches hidden in the Liberian countryside and are waiting to see how the new administration proceeds. Resolving these problems is necessary to move forward. It will require international donor commitment, the development of Liberia's security services, and the cooperation of Liberia's people.

Addressing the people's day-to-day concerns about personal safety and crime can stimulate practical cooperation and improve intelligence collection. Improved intelligence, in turn, can aid in learning about, and breaking up, potential insurgencies and other armed groups. When prevention is unsuccessful, the Liberian security forces must be able to defeat armed internal opposition, from gangs to rebellion. The danger of organized internal enemies demands security forces that are capable of fighting and winning military encounters, not just enforcing the law. If, however, Liberia's police are not careful when applying force, popular support and cooperation will not be forthcoming. In sum, Liberia needs a mix of internal security forces that can defeat organized internal enemies, in combat if necessary, while also providing public safety with restraint.

The third security element is preparing for external defense. This includes protecting Liberia's borders and responding to foreign aggression. The region has a history of states harboring and supporting groups hostile to their neighbors. Just as likely is the danger that neighboring states cannot control their borders with Liberia. For this reason, the Liberian government needs to be aware of movements across its borders and security developments in adjacent states. Good surveillance and other intelligence collection can bolster security from outside threats without a huge army. The capabilities and disposition of Liberia's army need not and should not give any of its neighbors cause for concern.

These considerations are summarized in Figure 2.1, which shows how core security functions flow from a national security concept.

Figure 2.1
Security Concept and Functions

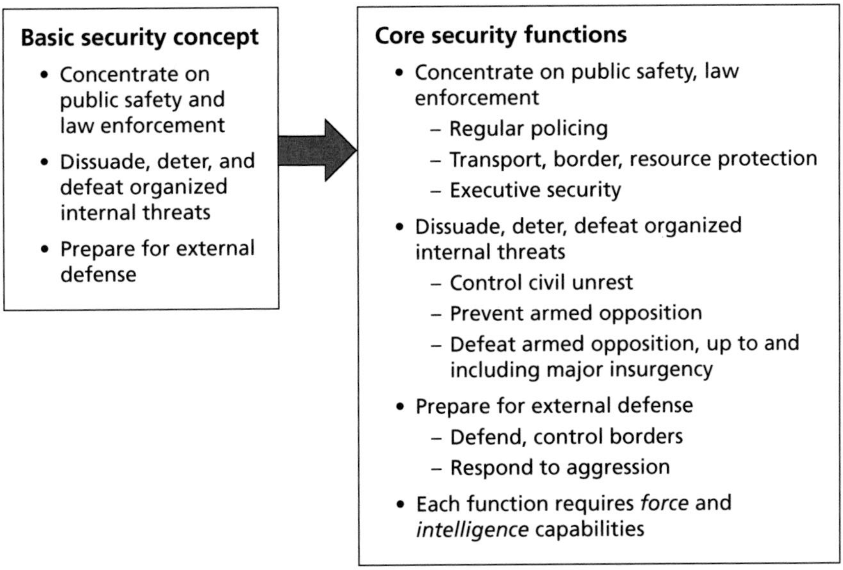

Basic security concept

- Concentrate on public safety and law enforcement
- Dissuade, deter, and defeat organized internal threats
- Prepare for external defense

Core security functions

- Concentrate on public safety, law enforcement
  - Regular policing
  - Transport, border, resource protection
  - Executive security
- Dissuade, deter, defeat organized internal threats
  - Control civil unrest
  - Prevent armed opposition
  - Defeat armed opposition, up to and including major insurgency
- Prepare for external defense
  - Defend, control borders
  - Respond to aggression
- Each function requires *force* and *intelligence* capabilities

CHAPTER THREE

Criteria and Principles

Criteria for Assessment

In a region with a precarious peace, the new Liberian government inherited a corrupt, bloated, incompetent, and unsustainable security sector. Past Liberian administrations, particularly Charles Taylor's, filled the security organizations with individuals selected for their loyalty or political desirability rather than their competency. When concerns arose over the loyalty of one security organization, Liberian leaders, rather than resolving the issue, created another organization to serve the same function and to control the original one. For example, upon gaining power Taylor not only replaced more than 2,500 AFL officers (including almost all of the remaining Krahn officers) with personnel from his NPFL, he also created Anti-Terrorist Units (ATUs), which were essentially a private army whose number eventually exceeded that of the AFL.[1] Liberia's national transitional government added management and personnel to the existing security organizations from the various factions in a political compromise, thus perpetuating the problem.

In order to succeed, Liberia must overcome the legacy of a corrupt, bloated, incompetent, and unsustainable security sector, and build a new security sector designed to meet four basic criteria:

- coherence
- legitimacy

---

[1] Pham (2004).

- effectiveness
- affordability.

The first criterion for Liberia's security sector is *coherence*. In the past, security agencies were created at the whim of Liberian leaders who altered the security sector to maintain power—resulting in a complicated, bloated, and incoherent security architecture inappropriate for a democratic regime. Because Liberia's internal security and external security are intertwined, a holistic view of the security sector is essential for the provision of any of the individual components of security. This includes not only the main building blocks of Liberia's new security sector (the AFL and the LNP), but also related ministries, specialized security services, and the judicial and court systems.

Currently, important and sensible efforts are being made in building various security components. These and future security sector transformation efforts must be coordinated to ensure that the overall security sector structure is appropriate for Liberia's internal and external security needs. For example, the size and mandate of the police force cannot be determined in isolation, but must be developed in conjunction with the role of the AFL. The roles and relationships among different forces must be clear, agreed upon, and monitored. Furthermore, the capabilities and resources must reflect an overarching architecture and set of priorities, as discussed in further detail in Chapter Four.

Coherence also demands a seamless connection between ends and means. It must be possible to trace resource requirements to real operating and investment needs. In turn, these must yield essential security capabilities. Setting capability requirements must flow from important operational demands and a national security strategy. This entire chain of analysis must originate with an objective assessment of the security environment and be publicly visible and comprehensible. Coherence underpins the other three elements: legitimacy, effectiveness, and affordability.

The second criterion for the security sector is *legitimacy*—not only in structure and oversight but also in the Liberian people's perceptions. Organizations such as Taylor's ATUs had questionable legitimacy because they were formed not to protect the security of the state or the Liberian people but rather to allow those in power to remain there. Past favoritism toward individuals from specific counties, ethnic groups, and political parties undermined the legitimacy of the security forces as organizations. The security forces were used to reward those loyal to the holders of power with employment and to provide protection for individuals who supported the government.

In the past, Liberian security forces were concentrated, either de facto or de jure, under the President, both to ensure their loyalty and to permit their unchecked use. A legitimate security sector requires that forces be under the control of the elected leadership, but without a concentration of power in any one person or agency and without the use of security forces for partisan ends. Positions in the security forces should not be utilized as rewards or patronage, but should be distributed on a competitive basis and be broadly representative of the general population.

Liberian security services must earn the trust of the general population through their actions and through their stewardship of resources. The security services must be responsive to the needs of the general population, not focused solely on the needs of individuals serving in government positions. Legitimacy can strengthen effectiveness; a population that trusts the security apparatus will cooperate willingly with it and provide information to it. Security services will then be able to manage disputes, thus encouraging respect for the security personnel.

The third criterion is *effectiveness*. If the security forces are not able to provide public safety, enforce laws, and protect against internal and external attacks, the security of Liberia and the Liberian people will be constantly at risk. Effectiveness has three main components: (a) qualified, professional, and well-trained personnel; (b) clearly defined mandates and coordination mechanisms; and (c) decisionmakers' access to accurate intelligence, objective analysis, and sound counsel from multiple and appropriate sources.

In the past, the deep-seated system of patronage for positions in the security sector made the security forces much less effective. Government leaders had little education or professional training. They invested little in developing professional capabilities for effective security forces. Liberia's security forces must be professional and of the highest quality that can reasonably be achieved. Training, capacity-building, and mentoring must occur at all levels of the security forces.

These activities should address issues that have historically hindered the effectiveness of the security forces. For example, drug use among former security personnel is reportedly widespread.[2] Efforts to create professional and quality security personnel must address such problems. Effectiveness entails not only professional, capable, and well-trained forces but also sound management and efficiency.

Effectiveness requires clarity about the roles, activities, and organization of the agencies. Overlapping mandates and ill-defined functions facilitate inefficiency, corruption, and abuse of power, as is evident from Liberia's history. Trust and collaboration within and between the security forces—up and down the ranks, among units or divisions, and among services—are essential. Clear chains of command must be understood and adhered to at all levels. Such trust and collaboration depend upon clear definitions of and common agreements on the mandates of each agency. Mechanisms for coordination and joint operation should be well defined; regularly exercised; and consistently used, monitored, and improved.

If the security sector is to be effective, decisionmakers must have access to unvarnished professional intelligence, security, and military advice. In the past, security forces were frequently politically motivated; they could not be trusted to provide accurate, unbiased, and essential security information. There was no well-functioning mechanism for coordinating reports from the various security agencies. Rather, each individual agency reported directly to the President, who then assessed the various reports himself. This not only perpetuated divisions among the security forces and incentives not to cooperate, but also significantly impaired the government's ability to make sound security decisions. An effective security sector requires open communication and high-quality, reliable security information and advice.

The fourth criterion for security sector is *affordability*, which is crucial in a state of Liberia's limited means. The creation of security forces that Liberia cannot afford to maintain and operate is a recipe for

---

[2]   Discussions with civil society members in Monrovia, March 2006.

ineffectiveness, corruption, and ultimately insecurity. As was the case with the AFL and LNP under Taylor, if the government is unable to pay the salaries of security personnel, those personnel will utilize their positions and power to extract money from the local population. It is commonly said that Monrovia's taxi drivers pay the salaries of the police officers.[3] Although bribery is always problematic, an inability to financially maintain the security sector could escalate into a return to civil conflict. A well-trained and armed military that does not receive its wages has the capability and the motivation to overthrow the government.

Although mismanagement of government revenue and corruption significantly hindered the payment of civil servants and military personnel in the past, bloated government payrolls also contributed to the lack of affordability (and legitimacy) of Liberia's security sector. The "right-sizing" of security agencies and reduction in overlapping organizations and functions will eliminate inefficiencies and reduce waste. These efficiencies and economies should be exploited. Resources should be allocated to finance the highest priorities and most important roles and missions.

The control and use of government funds, investments, and contracts must be transparent and civil servants must be held accountable. Doe's decaying mansion in Grand Gedeh County, with the Liberian flag painted on the bottom of its swimming pool, is a physical reminder of the widespread corruption and mismanagement of government funds of the past. Liberia's very limited budget revenues undermine the importance of an affordable and cost-effective security sector. This implies using government revenues to pay salaries for competent, necessary staff. The procedures by which government funds are collected, held, budgeted, obligated, and disbursed must be transparent and well-documented. The overall financial reforms to which Liberia and its international financial supporters are committed must be extended to the security sector.

Affordability must be viewed in terms of both operating costs and investment costs. Operating costs are largely a function of the number

---

[3]   Meeting with civil society members, March 2006.

of personnel, pay being a major factor. Consequently, personnel costs for a police unit of the same size as a highly capable military unit are roughly the same. However, the capital costs of military units are considerably more than those of a similarly sized police unit because military equipment is so much more expensive and so much more equipment is needed. Liberia and its partners should look at capital costs in light of the ensuing operating and maintenance costs, since Liberia is more likely to obtain external financial support for investments than long-term financial support to cover operating costs.

We use these four criteria to analyze the force-structure and institutional options and issues in the chapters that follow. These criteria should also be used by the Liberian government as it makes decisions affecting the security sector in the future.

Principles

Because Liberia's security situation has been so troubled and remains unsettled, the government should establish, communicate, and live by a set of principles to guide strategy, priorities, and conduct. The following principles take into account the current internal and external security environment, past abuses of security forces and the government, and future needs:

- *Adaptability for a fluid external environment.* Although the Mano River Basin region is currently relatively stable, it may not remain so. History indicates that the security situation in this region can deteriorate rapidly. For example, changes in Guinean politics could erupt in conflict that could spill across Liberia's borders or lead to the recruitment of Liberia's ex-fighters.
- *Flexibility in the context of a dynamic internal security environment.* The security sector cannot be developed as if the current levels of internal and external threats will remain constant; rather, it must be able to quickly adjust capabilities in response to a changing threat.

- *Accountability to and trust of the people.* Liberians remember the past abuses and deficiencies of the security sector. A high concentration of power in one individual or agency, as well as a lack of civilian oversight, could lead to a repeat of Liberian history. The security services should be under the control of the elected leadership and must be broadly representative of the Liberian population. The security services will need to earn the trust of the people—both through their behavior and through their stewardship of resources.

- *Quality as the key to effectiveness.* The new security sector must distinguish itself from its predecessors by its competence and capability. Because most potential adversaries are likely to depend on ill-equipped and ill-disciplined fighters, high-quality Liberian forces should have a decisive advantage. The presence of UNMIL provides time for Liberia to develop such high-quality, professional security forces. Training alone is insufficient, however, and oversight of the security forces must include monitoring, mentoring, and management. It is unrealistic to train new Liberian police officers and expect that this alone will address the issue of corruption. There needs to be significant monitoring of the police, with clear consequences for unacceptable actions. Quality also requires management not only of the forces, but also of the resources. Security sector leadership must buy into the tenets of the security sector transformation and have strong management capabilities.

- *Depoliticizaton of security forces.* In the new Liberia, the security forces should not be beholden to any particular party or individual in power but should serve the state and its people.

- *Distribution of power.* Excessive centralization of armed and intelligence power is no less dangerous than extreme diffusion. Too high a concentration of power facilitates oppression and has been the source of coups and internal conflict throughout the region and the world. The government should develop regional and local capacity rather than continuing to place the direct management of routine operations and daily activities in Monrovia.

- *Optimization of resource allocation.* Liberia's security sector will not operate effectively if it is not affordable. Dismissing redundant employees and exploiting contracting and organizational efficiencies will be necessary if Liberia is to be able to afford capable forces. Resources need to be allocated so as to correspond to the government's highest priorities and in accordance with the roles and missions assigned to each force or agency. The government will need to develop the ability to assess trade-offs between investments and current operating expenditures. These trade-offs include investing in personnel and affordable, but high-leverage, technology. Decisions must be based on careful analysis of the best balance between current operations and investment.
- *Transparency and accountability in financing.* Budgets and programs must be fiscally realistic, tightly drawn, monitored, and altered only with proper authority. As the administration of President Johnson Sirleaf transforms Liberia's security sector, the control and use of funds and contracts must be transparent and accountable to the elected civilian authority.

CHAPTER FOUR

Forces

Building Blocks, Roles, and Missions

The building blocks of Liberia's new security sector are the forces that enforce laws, provide for public safety, and protect the nation. The criteria of coherence, legitimacy, effectiveness, and affordability cannot be met unless these building blocks are each well designed and fit together. The specific sizes, capabilities, roles, and relationships of these forces must be linked to the assessment of Liberia's security environment, and their adequacy must be tested against the integrated security concept and core security functions.

The largest and most crucial components of Liberia's security sector are the Liberian National Police (LNP) and the Armed Forces of Liberia (AFL). The former should be the country's main internal security force; the latter should embody the country's main capabilities for military combat.[1] The size and capabilities of the LNP and AFL largely determine the effectiveness, cost, and thus the cost-effectiveness of Liberia's security sector. Their roles and missions and the relationship between them will largely determine how the new state provides security. Lack of clarity on missions risks duplication or gaps in capabilities, political contention, and operational failure.

- *The primary missions of the LNP are (a) to prevent and fight crime and (b) to maintain public safety.* These missions call for a light but

---

[1] Chapter Five looks at how ancillary forces, such as specialized police, the Coast Guard, and the Special Security Service relate to these two main building blocks.

sizable, community-friendly police force that can earn the confidence and cooperation of the Liberian people. Anticipating occasional civil disorder, the LNP should also have a branch capable of riot control—e.g., the police support unit (PSU).

- *The primary missions of the AFL are (a) to safeguard the country against possible external threats and (b) to support internal security forces in defeating any insurgency or other internal threat for which Liberia's internal security forces prove inadequate on their own.* At present, nonstate external and internal threats are more likely than threats from neighboring states. The size of the AFL is less important than that it be superior in quality and capability to foreseeable threats.

The missions of subunits of these main forces and of other ancillary forces must complement these primary missions. The complementarity of all Liberian security forces is key not only to their coherence but also to their effectiveness and affordability.

The method used here to analyze and select the most cost-effective combination of forces is to identify, assess, cost, and compare distinctly different options. Before presenting this analysis, it is useful to consider in very broad terms an architecture of capabilities.

Capabilities Architecture

Liberia's capabilities architecture should respond to a security concept whereby (a) public safety and law enforcement are immediate concerns, (b) the appearance of organized armed internal opposition must be anticipated and prevented, and (c) future external threats that may arise without long warning cannot be excluded.

Even with foreign assistance, Liberia's economy does not permit large forces. Moreover, increasing the size of planned Liberian security forces is unlikely to be the right way to yield the most security for each additional dollar spent. The key to cost-effectiveness for Liberia's security forces is to have complementary capabilities that cover the core security functions, possess the right qualities, and can be used flexibly.

The Liberian forces already being built with the help of the UN and the United States are consistent with these standards and are necessary. Analysis of possible operational contingencies suggests a need for an additional capability that would complete and tie together currently planned capabilities: a small mobile unit of the LNP that can perform either in a law-enforcement mode or in combat.

Such an LNP unit—a quick-response police unit (QRPU)—would complement the regular police. Unlike the police support unit, which is meant to deal with civil unrest (e.g., riot control), the QRPU would be capable of defeating organized armed threats. One of the most acute security dangers that Liberia faces is armed opposition forces—"proto-insurgencies"—that extend beyond the capabilities, law-enforcement mission, and normal training of regular police yet do not warrant the domestic use of the army. A small, well-prepared QRPU could help dissuade internal enemies of democratic Liberia from organizing armed forces. It could also operate jointly with the AFL in the event of a full-blown insurgency that might require the intervention of the AFL, or even against external threats, making it a "swing" force.

Qualified personnel from the regular LNP and PSU should be the source of police for the QPRU, though they would need additional training for light-combat missions. In turn, QRPU personnel as well as other police personnel assigned to the full range of units and functions could rotate back into the rest of the police force. Such rotations would add to the effectiveness, cohesion, and interoperability of the entire LNP and also ensure that the QRPU is organically tied to the regular LNP and under the firm control of its chief.

There are important arguments for treating the QRPU as a branch of the LNP instead of as part of the AFL or as a separate third force:

- Alignment with the police reflects the greater likelihood of use in support of and collaboration with the police in addressing internal security threats.
- Insofar as the QRPU operates in a law enforcement mode, its doctrine and training should be that of the LNP.

- Alignment with the army would require the government of Liberia to call upon the Ministry of Defense (MoD) and the army to intervene regularly in domestic security, which is not ideal politically and presents the added problem of dual command and control of forces and operations for internal contingencies.
- An independent third force would risk confusion and possibly contention both in command and control and during operations. It might not work well with either the LNP or the AFL.

In addition to being flexible, a quick-response police unit must be mobile. Although the rainy season and poor roads make mobility difficult, Liberia's geography and potential internal and external threats make it important. Precisely because it is impractical to make all Liberian security forces mobile, it important that some be. In the next few years, UNMIL's own quick-response force can provide mobility. In the longer term, Liberia and its supporters must consider their own options. Leaving large areas of the country and segments of the border beyond the reach of security forces is not tenable from the perspective of national security, but building large forces distributed throughout the country is not financially supportable. New roads may eventually ameliorate this problem, but building roads takes time. Even with adequate roads, rotary-wing air assets, provided by an international partner, can lend crucial operational advantages not only in fast transport but also in surveillance of areas where forest cover does not preclude it. (The accessibility-mobility problem is addressed further in the section below entitled "Testing Force Plans Against Potential Threats.")

As shown in Figure 4.1, the inclusion of a small quick-response unit of the LNP, able to operate in both law-enforcement and combat modes, would permit a complementary architecture responsive to the three-part security concept presented above.

For a country such as Liberia that faces a complex and fluid security environment and that must economize, synergy among capabilities, flexibility in their use, and the ability to respond swiftly and appropriately to unforeseen threats is crucial. This architecture provides for five alternative basic configurations:

Figure 4.1
Capabilities Architecture

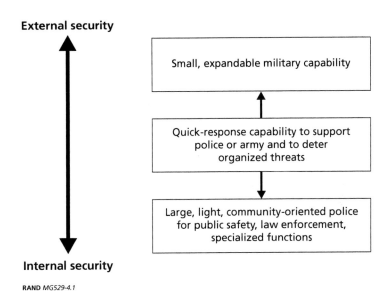

External security

Small, expandable military capability

Quick-response capability to support
police or army and to deter
organized threats

Large, light, community-oriented police
for public safety, law enforcement,
specialized functions

Internal security

RAND *MG529-4.1*

- Regular police (including PSU) acting autonomously
- Regular police acting with support from quick-response police
- Army acting autonomously
- Army reinforcing or reinforced by quick-response police
- Quick-response police used to confront organized internal threats.

These configurations offer flexibility, appropriateness, and responsiveness in an economical way, thus reducing the need for large forces.

Force-Structure Options

Employing this architecture, we examine three primary force options. As shown in Table 4.1, the options are differentiated by:

Table 4.1
Force Options

Option	Regular LNP	AFL	Coast Guard	QRPU
1	Small	Small	No	No
2	Large	Large	Yes	No
3	Medium	Small	Yes	Yes

- size of the LNP and AFL
- inclusion of a quick-response LNP unit
- inclusion of a Coast Guard.

Under this design, the QRPU would consist of approximately 750 personnel organized into three action companies: command, maintenance, and other support. Its equipment would be sufficient to defeat the kinds of armed internal opposition force described above. It would have organic road mobility but would be provided with air mobility by one of Liberia's international partners (e.g., UNMIL, at least at first).

Coast Guard patrol vessels are expensive to purchase and maintain. The skilled crews needed to operate and maintain these craft will be able to demand higher wages than entry-level security personnel. The Coast Guard needs to be sized so that it is affordable as well as functional. The force envisioned here consists of eight vessels: four 32-foot and four 28-foot craft.[2] It would employ 350 personnel. Such a force would be able to patrol significant lengths of the coast on a daily basis. It would not, however, be able to adequately patrol Liberia's 200-mile economic zone or to defend the country from a naval threat of any significance.

The three options displayed in Figure 4.2 are shown by size, measured in numbers of personnel or "end strength." However, we stress that end strength, in and of itself, is a poor measure of capability because it says nothing about training, equipment, quality of troops and leadership, or mission design. Also included for the sake of comparison is the current Liberian force structure.

---

[2]   Drawn from prior analysis by MPRI.

Figure 4.2
Force Size Options

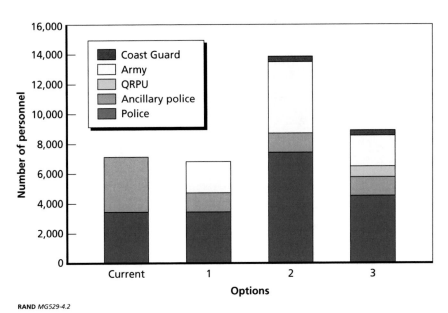

Effectiveness, Costs, and Cost-Effectiveness

In terms of effectiveness, Option 1 is unlikely to be sufficient for ensuring public safety, for protecting against organized internal threats, and for coastal security. It does not satisfy the basic requirements of the security concept, does not afford the flexibility and other advantages of the suggested architecture, and may not fulfill core security functions.

Although Option 2 has large police and army forces, their size alone does not ensure a capability to protect against threats, particularly those that arise suddenly. Option 2 could meet straightforward law enforcement and known external defense needs, as well as coastal security. However, given the lack of a quick-response police unit, Option 2 is potentially inadequate against organized internal threats without heavy reliance on domestic intervention by the army.

In contrast, Option 3 matches up well with the overarching security concept and fits the proposed capabilities architecture. It should enable Liberia to fulfill its core security functions.

To balance the guiding principles of effectiveness and affordability, it is important to consider both the operating and capital costs of the various options, depicted in Figures 4.3 and 4.4.[3]

Although Option 1 is the least expensive in terms of annual operating cost (about $17 million), it does not meet Liberia's security needs. The operating costs of Option 2 are $18 million more a year than those of Option 1. Although Option 2 is the most expensive option on an operating basis, it is nevertheless unlikely to be adequate to counter organized internal threats without heavy reliance on army intervention. Option 3, which should address Liberia's security needs, costs

Figure 4.3
Operating Costs of Options

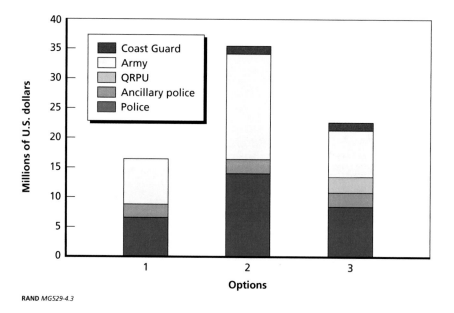

RAND MG529-4.3

---

[3] These cost estimates do not include expenses for leasing helicopters for airlift and surveillance.

Figure 4.4
Capital Costs of Options

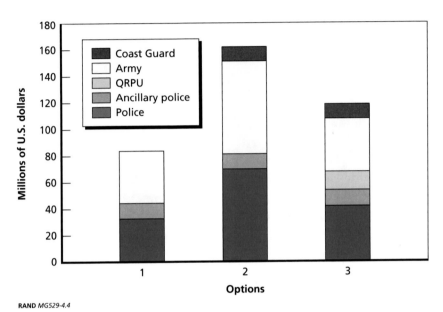

RAND MG529-4.4

approximately $22 million to operate annually, or $6 million more per year than Option 1. Option 3 costs approximately $119 million to build (capital costs). This is $35 million more than the capital costs of Option 1 but $43 million less than those of Option 2.

The return on the additional investment for Option 3 compared to Option 1 is significant: *a lower security risk at an affordable increase in operating cost.* Option 2, which entails an expansion of the police and army, is less effective and considerably more costly than Option 3 to build and operate. Option 3 is not only the most effective option in terms of security objectives but also the most cost-effective.

Internal and External Balance

Balance in terms of strength among the several Liberian security forces and among the ministries to which they report is also an important objective. As legitimacy, civilian control, and National Security

Council-based collective oversight become rooted, concerns about balance should decline. However, as long as there is a danger of political instability, balanced forces are important.

- In Option 2, the AFL would have vastly greater capabilities than all other Liberian security forces combined and would alone be capable of combat operations.
- Option 3 would provide better balance by virtue of the relative sizes of the force and the inclusion of the QRPU in the LNP.

A gross discrepancy in military forces between Liberia and neighboring states would be inadvisable, even under current, relatively unthreatening conditions, if only because it would cause an imbalance of burden to secure common borders. Of nearby nations:

- Sierra Leone's army is 10,500 strong[4]
- Guinea's is 12,000[5]
- Côte d'Ivoire's is 20,000[6]
- Ghana's is 7,000[7]
- Senegal's is 17,000.[8]

UNMIL, at peak strength, was about 17,500 strong. To put these numbers in perspective, the average number of citizens per solider in West Africa is 1,285; the number of citizens per soldier is 611 in Option 2 and 1,420 in Option 3. The average square miles per soldier in West Africa is 16.45; the square miles per soldier is 9.06 in Option 2 and 21.06 in Option 3.[9]

---

[4] Meetings in Sierra Leone, March 2006.

[5] U.S. Department of State (2006b).

[6] U.S. Department of State (2006a).

[7] International Institute of Strategic Studies (2006).

[8] U.S. Department of State (2006c)

[9] International Institute of Strategic Studies (2006) and CIA (2007).

These comparisons raise the question of whether the planned AFL, as shown in Option 3, is sufficiently large. We argue that it would be large enough because

- UNMIL will not depart Liberia for some years to come. The LNP is at this stage Liberia's highest priority.
- The LNP's QRPU, once built, will be available to augment the AFL
- the quality of the AFL is more important than its size. With the time provided by UNMIL's presence, Liberia and the United States should be able to concentrate on making the AFL a high-quality force
- with an adequate LNP, including a QRPU, internal security demands do not merit a larger army.

In any case, given the unpredictable nature of Liberia's external security environment, expansion of the AFL would be possible and may be necessary at a later time. If that proves to be the case, Liberia's force structure can be adapted. In sum, it is neither necessary nor helpful to decide now how much larger than 2,000 troops—if any larger at all—the AFL should be.

Testing Force Plans Against Potential Threats

To test the adequacy of the force structure options, Figure 4.5 depicts a spectrum of threats from least-dangerous and most-likely threats to least-likely and most-dangerous. It shows how key security forces and services align with threats along this spectrum. "P" denotes principal responsibility; "S" denotes a possible supporting role. The gray-shaded area identifies the higher-end dangers that seem plausible over the next few years.

Figure 4.5
Spectrum of Dangers and Capabilities

	Day-to-day crime	Border or coastal crime	Organized crime	Low-grade civil unrest	Violent civil unrest	Armed gangs	Minor insurgency	Major insurgency	Non-state incursions	External aggression
Regular LNP	P	S	P	P	S	S				
Immigration		P								
Coast Guard		P								
PSU				S	P	S				
QRPU					S	P	P	S	S	S
AFL								P	P	P

P = Principal role
S = Supporting role

RAND *MG529-4.5*

We find the following:

- The figure covers the entire spectrum of plausible threats.
- The critical gray area from violent unrest to major insurgency is adequately covered.
- Every force has at least one principal responsibility.
- In some cases, multiple forces could, and should, be involved. In each of these cases, the force that has principal responsibility should be clearly identified.
- Any redundancy implied by the relevance of more than one force to any given threat is by design and not duplicative.
- In the absence of a QRPU, the AFL would have to intervene domestically against even minor organized threats.

The schematic does not capture all the nuances and ambiguities that may attend actual contingencies, especially in the gray area. But it does offer both a test of adequacy and a presumptive indication of responsibility along the spectrum.

Liberia has few roads, and they are in poor condition. Many are impassable during the rainy season. Even after the road system has been improved, washed-out roads will render parts of the country unreachable by vehicles. For UNMIL and future Liberian forces, such limitations on surface mobility can be partially overcome by rotary-air mobility (provided by a foreign partner for Liberian forces). This lack of mobility introduces a risk that armed internal opposition groups could organize, operate, and even seize control of areas inaccessible to state security forces.

Awareness of this risk is the first step toward reducing it. The government of Liberia should ensure surveillance of areas that are difficult to access by any practical means, including police and other human contacts. Early intelligence of a threat-in-the-making is critical to permit small-scale but effective preemptive action by the QPRU. Failing that, isolation of such areas may render organized threats less harmful and possibly less viable. If it is impossible to destroy armed opposition groups in temporarily inaccessible areas, use of larger road-mobile QRPU or AFL forces is still an option when the roads are open.

An internal enemy force that must remain confined to an area inaccessible to state security forces—though not to be tolerated insofar as is practical—will pose a limited danger to the Liberian nation.

International Forces and Integrated Force Plans

Liberia needs an international presence to serve as a deterrent until it has created satisfactory security forces. The presence of UNMIL or a similar international force[10] gives Liberia time in which to build professional, well-trained, and high-quality security forces. As the Liberian

---

[10] For purposes of simplicity, we refer to the international force as UNMIL. Over the course of time, the name or even the source of the international forces in Liberia may change.

security forces develop and become functional,[11] the international presence can be reduced, as shown in Figure 4.6.[12]

This figure can be read as follows:

- T1: Once the building of new Liberian security forces is well under way and units have begun to deploy, UNMIL could begin to draw down its least-capable, or least-relevant, forces.
- T2: By the time Liberian forces are near their full planned strength, UNMIL could be down to perhaps half its current size.
- From T1 to T2: UNMIL should tailor its forces to complement Liberian forces.

Figure 4.6
Integrated Force Plan[13]

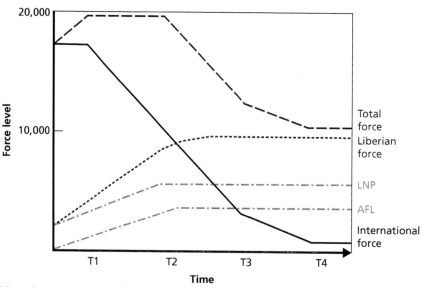

RAND MG529-4.6

---

[11] Under current plans, it is anticipated that all AFL soldiers will have completed basic training by the end of 2008 and that all police officers will have completed training in 2007.

[12] We stress that the UNMIL force levels depicted are only illustrative and do not represent the views of any of the parties. The Liberian force levels are based on current plans.

[13] The present LNP levels are based on personnel who have undergone UNMIL training.

Especially important would be
- UN police (UNPOL) elements, to continue to advise, mentor, and develop the Liberian police force capabilities
- the UNMIL quick-response force
- rotary-wing airlift and surveillance
- Reduced regular land forces.
- T3: As Liberian forces gain in field experience and quality, UNMIL could reduce its force size to perhaps 3,000 troops, concentrated on the key capabilities just described.
- T4: With Liberian forces able to secure the country largely on their own, UNMIL or some other international partner, or commercial provider, would nevertheless maintain small, critical "enabling" capabilities: advisors and modest air transport and surveillance to support Liberian forces.

Illustratively, based on current plans, and assuming favorable security conditions and Liberian force development continuing apace, T1 could be 2007, T2 could be 2010, T3 could be 2012, and T4 could be 2013.

Table 4.2 shows rough cost calculations for different UNMIL force levels.[14]

Although it is unclear that the UN will be prepared to maintain any presence beyond the time Liberian forces reach full strength, Liberia's history and regional environment strongly argue that a reduced and tailored presence could be needed at a drastically reduced cost for a few years thereafter to ensure conflict and chaos do not return.

Table 4.2
Projected Cost of Maintaining an International Presence in Liberia

Force Levels	9,000	3,000	500
Cost (millions of US$)	$460. 8	$51. 2	$25. 6

---

[14] Calculated on the basis of the average cost per soldier for UNMIL of $51,201 per year.

During the transition, command and control arrangements between UNMIL and Liberian security forces must be delineated and coordinated with great care from headquarters to the field.

In closing, it should be noted that whatever type of force Liberia selects, it will need to be flexible not only in its capabilities, but also in its size and structure. In coming years, as Liberia stands up its ministry of defense and its other security agencies are nationalized, a national capacity to assess the changing threat environment and to derive force requirements for it must be developed. Based on these assessments, Liberia may choose to alter force sizes and mixes in the future. The structures outlined here could easily be grown, shrunk, or otherwise adapted in line with such policy choices.

CHAPTER FIVE

Organizing Government

However well designed, Liberia's security forces will not be coherent, legitimate, effective, or affordable without a governance structure that also meets these criteria. That includes not only an elected government at the top but also the ministries and agencies that manage day-to-day administration and operations; the ways in which they interact with one another; and the regulations, rules, and laws that bind them as they enforce Liberia's laws and ensure its integrity.

Current Security Organizations

Liberia's current security sector is characterized by redundancy, inadequate control, and incoherence. The most obvious concern is the sheer number of structures that exist. The new government inherited no fewer than 15 separate agencies and structures tasked with a variety of security functions, some discrete and some overlapping (Table 5.1).

It is not simply the number of agencies that is cause for concern but also the redundancy and ambiguity concerning their functions and roles. All of these agencies, with the exception of the Ministry of Defense, have the authority to arrest and detain individuals. The National Security Agency (NSA), the Ministry of National Security (MNS), the Liberian National Police, the National Bureau of Investigation, the Ministry of Defense, and the Special Secret Service all collect intelligence, including criminal intelligence, political intelligence, and—in the case of the MNS and NSA—foreign and national security intelligence.

Table 5.1
Current Security Organizations

Defense	Executive Protection	Intelligence	Policing
Ministry of Defense		Bureau of Immigration and Naturalization (BIN)	
		Drug Enforcement Agency (DEA)	
	Special Security Service (SSS)		Customs—Financial Security Monitoring Division (FSD
		Ministry of National Security (MNS)	Forest Development Authority Police (FP)
		Liberia National Police (LNP)	
		National Security Agency (NSA)	Liberia Petroleum Refining Company Security Force (LPRC)
			Liberia Seaport Police (LSP)
			Liberia Telecommunications Corporation Plant Protection Force
			Monrovia City Police (MCP) also known as Department of Traffic and Public Safety
		National Bureau of Investigation (NBI)	
			Roberts International Airport Base Safety (RIA)

RAND *MG529-T-5.1*

A variety of security organs, performing a range of functions, even overlapping ones, is not in and of itself unacceptable. Wealthy countries often have complex and even inefficient structures to provide insurance and redundancy. However, such an approach is expensive and difficult to manage. As U.S. experience has shown, multiple intelligence agencies can lead to confusion and a failure to share information. For a country the size of Liberia, neither the multiple services nor the required management structure are affordable or justifiable.

Multiple security services are also a breeding ground for politicization and corruption, as officials in charge of various structures may come to view them as their private armies. The loyalties of uniformed personnel become personalized, rather than focused on the state. This is particularly dangerous if those individuals also represent factions vying for power, creating a danger for a coup or civil war. Forces may also compete for the favor of the government, as intelligence and security forces did in Liberia under Charles Taylor. This diverts forces from their core function of ensuring the security of the public and the state. The different agencies also tend to seek to raise funds themselves, whether authorized or not, promoting corruption and diverting them from their missions.

A welter of security services confuses both Liberians and visitors. Currently, many government officials, civil servants, and the Liberian citizens do not understand the roles and powers of the various forces. Travelers crossing the border into Liberia may encounter officials from the Bureau of Immigration and Naturalization (BIN), Customs and Excise, the Liberian National Police, the Ministry of National Security, the National Security Agency, Defense Intelligence, the Ministry of Health and the Ministry of Agriculture. The visitor has a difficult time understanding the purposes and powers of these agencies, and the confusion leaves room for abuse.

In sum, the current system facilitates corruption, is an inefficient use of state resources, and, if retained, could undermine the success of Liberian democracy. A small number of organizations, with clear mandates and minimal overlap, is needed.

Streamlining the various structures and agencies will not a simple task because of the variety of legitimate functions that must be performed. Even as too much diffusion of power is dangerous, so is too much concentration. Although Liberia's police and other security structures are not armed at present, they may be in the future. Concentrating all or most of the country's armed force under a single ministry gives that minister excessive power and perhaps the temptation to wield that power for personal political gain. A balance must be struck between differentiating functions and limiting overlap.

Another problem plaguing Liberia's current security framework is inefficient and inadequate oversight of security forces. Substantial efforts to train and rebuild the security forces are under way, but they will be undermined without appropriate oversight and monitoring. Effective oversight is a question of lines of command, hiring and firing authorities, and reporting chains, all issues that are confused and confusing in Liberia's extant agencies. For example, the Ministry of Justice (MoJ) has de jure oversight of the LNP and NSA but little authority over them in practice. Under the old regime, these institutions easily and consistently bypassed the MoJ and reported directly to the President. These institutions occasionally worked closely with the MoJ, but the more usual practice was to bypass the Minister and report directly to the President.

This highlights another challenge: centralized power within the office of the President. While appealing from the perspective of having a single voice and a single decisionmaker, too much authority concentrated in one individual can make everyday decisionmaking impossible, as everyone waits for a single extremely busy individual to make the most trivial decisions. Moreover, such centralized control has been abused by past holders of the presidency in Liberia as well as elsewhere. The high respect in which President Johnson Sirleaf is held is no guarantee against difficulties or concerns in future administrations. In most effective governments, the chief executive has ultimate oversight over security decisions but is not heavily involved in the daily management. This also helps to differentiate loyalty of security services to the person of the President from loyalty to the state.

The criteria of coherence, legitimacy, effectiveness, and affordability suggest that Liberia should have a new, properly distributed security sector institutional architecture that is clearly codified and communicated. Whichever force structure is selected, it is doomed to failure absent appropriate decisionmaking and command authority structures. The balance of this chapter lays out some options for alternative structures of national security organization and decisionmaking and evaluates the costs and benefits of each.

National Security Decisionmaking Body

The first considerations are how security policy is formed, how resources are allocated, and how crises are managed at the highest levels of government. On paper, Liberia presently has a Joint Security Committee chaired by the Minister of Justice, whose role is to oversee the performance of such functions. However, this structure has faltered in fulfilling these roles:

- It does not include several important views, such as those of the Ministries of Finance and Foreign Affairs.
- Actual authority over policy, resources, forces, and operations is unclear and limited.
- Ministers do not feel obliged to treat committee decisions as policy guidance.
- By placing leadership in the Minister of Justice, the President is removed and shielded from the security decisionmaking body. She or he can then make decisions without consulting that body. Without the authority of the President, it is unclear how the committee's decisions will be implemented. This not only further limits the committee's authority but also creates the potential that critical decisions could be taken without the President's involvement.

An alternative is to rely on a National Security Council, along a model similar to that employed by the United States. The U.S. National Security Council is chaired by the President. In Liberia, it should comprise the Minister of Defense, Minister of Justice, Minister of Finance, Minister of Foreign Affairs, and other ministers as required, on an issue-by-issue basis. Advisors to the National Security Council should include the chief of intelligence, the military chief, the police chief, and the Liberian National Security Advisor (LNSA). Moreover, during transition from UNMIL to Liberian forces, the President may choose to also invite, for example, the SRSG to take part in some NSC discussions to better inform debate.

The advantage of this option is that it could help ensure that the security decisionmaking process is authoritative, inclusive, and informed. Given the transparency and involvement of all relevant government agencies, this system, if used consistently, is hard to abuse and engenders collegiality and trust. The disadvantage of this option is the potential for indecisiveness if the President does not exercise effective leadership.

In some governments, decisions on security policy are made by consensus behind closed doors. Dissenting voices have the option to leave the government or to remain quiet in the aftermath of a decision. In others, decisions are made by majority rule. In both instances, the President can be the final arbiter and rule against the advice of her or his ministers—although she or he then faces the possibility of mass resignations of key national security staff. These decisionmaking mechanisms can be formally laid out or informally defined by each successive President and her or his cabinet. The options, however, should be well understood.

To be effective, an NSC system must reach deeper than the cabinet level. From subcabinet to working levels, interagency communication, including the participation of the armed and intelligence services, must be a way of life in the Liberian security sector. The complex nature of the security challenges faced, the need for coherence in implementation, and the demand for collaboration among different services in the field require continual interagency information-sharing, face-to-face meetings to plan and resolve issues, and preparation of options for decisions by the NSC itself. At every level, such a system will only work if ministries and individuals are open, trustful, trustworthy, and devoted to common, national objectives. That this mode is a challenge for even the most advanced and experienced states makes it no less important as a standard for Liberia. Although it will take a long time before Liberia achieves interministerial cooperation at every level, this is all the more reason to initiate it now. It is up to the President to set the standard and to display and insist on the cooperative habits needed to meet it.

The responsibilities and qualifications of the LNSA are key to an effective NSC system. The LNSA has three key general responsibilities:

- Orchestrating the NSC system at and below the cabinet level, in policymaking, priority-setting, crisis management, and implementation that cuts across ministries
- Ensuring that the President and the rest of the NSC receive objective analysis, options, and all points of view
- Fostering good direct ties among key ministries and agencies.

In Liberia's case, two other responsibilities are important in the early years of the new state:

- Monitoring the progress of security sector transformation
- Monitoring the quality of operational cooperation among the various security services.

The "ideal" National Security Advisor does not try to manage activities in ministries and services that fall within the responsibilities of those who lead them. While the National Security Advisor may have views on policies and priorities, those views should not interfere with the fair presentation of others' views and options. The LNSA must not come between the President and her or his ministers.

Although it is expected that each incoming President would appoint a new LNSA, it is important that the individual have broad respect for inclusiveness and objectivity, not simply the attention of the President.

National Military Command Authority

Although the Ministry of Defense has been in charge of the armed forces in theory, Liberian presidents have traditionally exerted direct control over the military in crisis, with the Minister of Defense having little independent authority. Although this arrangement has the advantages of decisionmaking clarity and speed in a crisis, it leaves the Defense Minister and the MoD with an ambiguous and weakened role. Liberia's history and that of many other African countries amply illustrate the potential for abuse of such a crisis command and control structure.

Many countries have chosen to reject such a system and maintain peacetime defense structures even in crisis and war. The decision to use force is the President's to make, as advised by her or his national security decisionmaking staff and advisors (e.g., the NSC, in Liberia's case). However, with regard to how forces are employed, although the President retains overall control and remains the commander-in-chief at all times, she or he delegates most decisions through a civilian Minister of Defense. This option is harder to abuse; it is also clear and aligns the crisis chain of command with the peacetime chain of command. The sole disadvantage is that there may be contention over the use of force and potential delay—a price worth paying in return for avoiding the pitfalls of absolute presidential authority over the use of military force.

Domestic Use of the Army

As explained in Chapter Four, the primary missions of the AFL are to safeguard the country from external threats and to assist internal security forces in defeating any insurgency or other internal threat that exceeds the capability of those internal security forces. Assuming the LNP includes a QRPU capable of low-scale combat operations, the need for the AFL to take forcible action inside the country should be less than if no such unit existed. Police primacy for domestic security is a critical element of a democratic society; even with its domestic security challenges, Liberia should be no different.[1] Military and police concepts of operation and rules of engagement are very different. The domestic use of military force can be traumatic to a society, and even the prospect can create anxiety about the military's purpose. Frequent

---

[1]   In this regard, the present plan to base the AFL near the capital while it is being built is not ideal for the longer term. Before any final decisions are made, further study of basing options is needed, taking into account security conditions around the country and along the borders, refined AFL concepts of operations, and mobility.

use of the military as a mechanism for civil authority, though common in many countries, is correlated with abuses, authoritarianism, and oppression.

Many countries do lay out procedures under which military forces may be used internally if police are unable to cope with the threat. The barriers to such action are usually designed to be high and to require top-level decisionmaking to temper the potential for abuse and to limit the situations in which military force can be used. Clear procedures and checks and balances are crucial. Liberia has several options for how these can be structured.

One option is that, prior to the use of Liberia's military forces in a domestic capacity, the LNP must request that the NSC issue an order for the Ministry of Defense to provide military support. Alternatively, the NSC can make the decision to call on the army to act whether or not the LNP makes a request. This is important for several reasons. First, the Ministry of Defense cannot act unless the LNP requests support or the NSC orders it. In either case, the army cannot act unless there is agreement on the part of the NSC, and thus the President, to authorize it to do so.

The role of the legislature is also important. In one option, if the NSC decides to issue an order for domestic use of the military, it should notify the legislature of its decision. The advantages of this option are that it can facilitate quick domestic use of the military when needed while maintaining government deliberation and accountability. On the other hand, the limited role of the legislature raises questions of legitimacy and political sustainability. Under this option, domestic use of the armed forces remains effectively an executive branch prerogative.

A second option also requires the LNP to request, or the NSC to order, the Ministry of Defense to employ the military domestically. In this case, however, the legislature, upon being notified, may call a halt to the military action (perhaps by a two-thirds vote). Because the onus is on the legislature to halt military action, rapid response remains possible. However, giving the legislature this power strengthens public accountability. If the legislature does not halt action, its acquiescence in the decision to use force is assured. The advantage of facilitating quick domestic use of the national military is that the decisionmaking burden is shared. However, the possibility that a decision to use the military would be reversed could leave a major domestic threat unchecked.

Given the probability of armed internal opposition and the need for strong deterrence and decisive actions, it may be prudent at this juncture to require only notification of the legislature. However, as both security and polities stabilize, this should be revisited.

Police and Policing Functions

Most of Liberia's myriad security services fall broadly under the banner of policing. Policing functions are an obvious starting point for streamlining and rationalizing security structures. The LNP is the core of Liberia's internal security organization. It has responsibility for national and local policing functions across the spectrum, and its reform has been undertaken with significant assistance from UNMIL and donor countries. At the time of this writing, Liberian police vetting and training is under way, although problems with effectiveness, redundancy, and retirement of members of the old LNP remain.[2] These problems will be discussed in Chapter Six.

In addition to the LNP, several ancillary forces exist. In many instances, their functions duplicate the LNP's law enforcement mission. One of these is the National Bureau of Investigation (NBI), whose functions include the investigation of major crimes (including homicide, robbery, arson, rape, forgery, theft of government property, and others) and other investigations, as assigned. The NBI under Charles Taylor also had a domestic intelligence role, which overlapped with the functions of the NSA and MNS. The Drug Enforcement Agency (DEA) also has functions that are identical to those of the LNP narcotics unit and the NBI, which also is supposed to have a counter-narcotics function. The DEA is reportedly staffed by police personnel. It also reportedly has some domestic intelligence functions. The Monrovia City Police is engaged in patrolling, traffic control, and crime preven-

---

[2]   Discussions in Monrovia, February 2006.

tion within the capital city, but the LNP performs these functions in Monrovia as well.[3]

Other existing services combine policing and nonpolicing roles. The Bureau of Immigration and Naturalization, Customs and Excise, and the Special Security Service (SSS) have very specific missions. In most countries, counterpart structures for these agencies are not part of the police reporting chain, although they do have some policing responsibilities.

The Bureau of Immigration and Naturalization is responsible for controlling the entry of individuals into Liberia. BIN needs arrest and detention authority so that unauthorized individuals, vehicles, or vessels can be stopped at the border.

In contrast, the primary function of Customs is to collect import and export duties, not to provide security. In the pursuit of its duties, Customs needs the authority to inspect individuals, vehicles, and vessels; to seize and hold contraband; and, potentially, to detain individuals suspected of smuggling, unless other law enforcement personnel are readily available to carry out that task.

Because Liberia does not have a Coast Guard, it is unable to prevent foreign fishing fleets from exploiting Liberian territorial waters or to keep individuals or contraband from being smuggled into or out of the country. A Coast Guard, even of modest size, would help the Liberian government to charge and collect fees for the use of Liberian territorial waters and to deter smuggling operations.

A key problem facing the effective use of a Coast Guard is corruption. Because Coast Guard ships operate autonomously, smugglers can offer bribes in exchange for permission to proceed with their activities. Fishing boat captains can pay off commanders to avoid paying fees. Although such corruption is difficult to combat, placing customs agents on Coast Guard craft provides another set of eyes. Constant radio contact and reporting on what vessels are sighted and where provides some control. Identification of ships prior to boarding permits commanders on shore to order the patrol craft to seize a ship before a bribe can be arranged. All fees and fines need to be paid to the Min-

---

[3]   Discussions in Monrovia, February 2006.

istry of Finance (MoF), not to the Coast Guard. The Coast Guard should be able to apprehend and detain (for short periods); it should not be permitted to levy and collect fines.

In order to support the Coast Guard capacity, a system similar to that instituted in Sierra Leone should be considered—perhaps in cooperation with Sierra Leone. Under that system, fishing boats are issued transmitters along with their fishing licenses. Sierra Leone leases aerial surveillance, which patrols to ensure that only boats with transmitters are fishing in patrol waters. If Liberia implements such a system, and interlopers are thus discovered, the Coast Guard could then be called to respond.

The Special Security Service, whose duty it is to protect the President and those designated by him or her for special protection, has a function that includes law enforcement but is primarily focused on personal protection—a very different mission. An effort to reform and restructure the SSS is under way, and should continue, with the support and help of international donors.

Finally, there are those structures whose duties are primarily security, with secondary law-enforcement roles. The Liberian Seaport Police are responsible for police and other functions at the ports, including crime prevention, response at the ports, searches of and control over entry of vehicles and individuals en route into or out of the port. The Forest Development Authority (FDA) Police are responsible for ensuring that forestry and conservation laws and regulations are enforced and that forests and forestry personnel are protected. Both of these institutions are financed by their parent agencies, from funds generated from the operation of the Port Authority and the Forest Development Authority, respectively. Their functions are a combination of policing, security, and entry and exit control.

The security and protection forces for the Telecommunications Company, Petroleum Refining Company, and Roberts International Airport are responsible for security, law, and order at these facilities, as well as airport passenger control.

Option 3 in Chapter Four proposes that Liberia retain medium-sized police forces, with ancillary capabilities to carry out the tasks for which the currently separate agencies are now responsible. This can be done in a number of ways. The key question here is which structures to incorporate into the LNP, and which to leave independent.

Most of the functions of the NBI (except any remaining domestic intelligence collection role), the DEA, and the Monrovia City Police are similar to LNP functions. The missions of these services already exist under relevant units of the LNP (or, in the case of the NBI and DEA intelligence roles, under other intelligence organizations). Keeping these organizations independent involves high continuing costs, significant difficulties in coordination, competition among organizations, and great potential for corruption. It also would require separate recruiting, training, and vetting operations for these organizations, as well as the creation of appropriate oversight bodies. Substantial savings and efficiencies could be realized by placing these responsibilities in the LNP (with intelligence missions being consolidated under the intelligence organization). The organizations themselves should not, however, be transferred en bloc into the LNP. Rather, existing personnel could be individually recruited and trained in accordance with LNP requirements, retired, or declared redundant. In this way, the existing plethora of independent police organizations could be consolidated within one agency, the LNP. Specialized functions would be performed by specialized bureaus within a single law enforcement agency, the LNP.

The BIN, Customs and Excise, the SSS, and the Coast Guard are probably best kept (or, in the case of the Coast Guard, developed) as separate agencies with their own reporting chains. In all cases, a comprehensive vetting, recruiting, and training effort, not unlike that of the LNP, must be undertaken. Such an effort is currently under way at the SSS, and is planned elsewhere. It must be implemented.

The primary mission of BIN is to monitor and secure Liberia's territorial boundaries. Because it needs to have arrest and detention authority, it is logical to keep BIN under the supervision of the Ministry of Justice.

The primary mission of Customs is to collect revenue, so Customs is probably best situated where it currently resides—inside the Ministry of Finance. Remaining inside the Ministry of Finance facilitates coordination with other tax units. It also makes it easier to change and improve the operations of Customs. For example, prepayment of customs duties at the moment of shipment rather than when the cargo reaches Liberia helps reduce corruption and better ensures the integrity of payments. In particular, if the Ministry of Finance obtains invoices electronically from the exporter, it can better monitor the actual value and composition of the shipments. As Customs shifts increasingly toward electronic payments and other means to better monitor imports, coordination with the Ministry of Finance will become more rather than less important.

The SSS would report to the office of the President. It requires different training than that of other law enforcement agencies and needs a different command structure. Like the Bureau of Immigration and Naturalization, the Coast Guard, once built, should report directly to the Ministry of Justice because it has primarily a policing mission, although that mission is different from the policing missions of the LNP.

The protection forces assigned to state-owned companies operate chiefly as security guards. They do not need the level of training required for police. The simplest option for these organizations is not to treat them as police units but to reconstitute them as security details financed by state-owned enterprises. The enterprises may choose to make these forces their own employees, convert them into separate units that would bid against private bidders for security contracts, or disband them as the LPRC has reportedly begun to do.[4] Security guards would not have arrest authority, but they should liaise with local police as needed—a link the LNP should facilitate. The Liberian government might choose to demand that these private security personnel undergo background checks and provide their forces with training. Moreover, under these options, the government may want to consider deploying additional police to areas of particular strategic concern at times of elevated threat.

Another option is to create a government-funded and -managed "police auxiliary" assigned to guard these facilities. Assigning security guard functions to a government agency introduces considerable inefficiencies. It is not clear who would pay for these forces, how salaries

---

[4]   Discussions in Monrovia, May 2006.

should be set, or what level of training would be necessary. A government-funded force also reduces or eliminates competitive pressures to provide security in a cost-effective manner. It would introduce a bureaucratic layer that, based on Liberia's past experience in this area, would likely introduce a great deal of inefficiency and could become a means for individuals to use their positions for personal material gain. A government-funded force would have the advantage of standardized training requirements and procedures. But unless it was carefully managed, it could create public confusion regarding who are police and who are the unarmed auxiliaries—a confusion that could be abused. Such an option would also likely be more expensive than a system strictly financed by the state-run companies themselves. At the same time, the creation of a large number of unskilled (and unarmed) jobs for this function by the government would be well received politically.

A third option would be to ask the police to serve as guards for these facilities as one of their rotational duties. This would be an expensive and inefficient use of law enforcement officers, creating a need for more trained police simply to fulfill these guard duties—which would not require the full range of police skills.

Airport security personnel should be responsible for checking passengers for weapons and explosive devices and for guarding the airport and aircraft. These responsibilities are primarily those of well-trained security guards, although they have a greater degree of responsibility. These individuals do not and should not replace customs and BIN agents at the airport; they would not have arrest authority. They would need to call on police or BIN agents to arrest a suspicious individual. As with other security guard forces, airport security should be managed and paid for by the airport.

Seaport and Forest Development security personnel represent a similar situation because somewhat more specialized knowledge is needed to fulfill those tasks, but the tasks remain primarily those of security guards. However, a combination of privatization and assignment of some key duties to police personnel should be possible in these cases, as well.

Specialized functions could exist as separate services under the LNP, with personnel assigned to one such function or another, with appropriate training. The LNP already has a number of specialized functions, including the PSU. Narcotics, economic crime, and other such areas can become similar specialties. This action would help decentralize authority and ensure the functional focus of these organizations. Consolidation with other police services of all of these functions will also help to combat corruption, ensuring that the same standards of conduct apply to all police.

An alternative to a number of separate specialized services under the LNP, with personnel more or less permanently assigned to them, is to rotate personnel through different functions, giving them appropriate training as needed.[5] The advantage of this approach is that it might reduce corruption because individuals would not be permitted to build up personal fiefdoms and contacts from whom they could demand bribes. It would also serve to ensure that officers engage in additional training during their careers. Over time, more and more police will have training in more and more functions, as they rotate through the gamut of police responsibilities. This will ensure that senior police are well prepared for a range of management roles and that qualified personnel are available when needed to backfill a variety of tasks—an important capacity-multiplier for a small country. The concentration of power under this option is considerable, however, and worthy of concern. Although the rotational system helps spread capacity, it also has some potential to erode functional specializations, although the bureau structure used by many law enforcement agencies around the globe would likely help prevent this erosion.

One final issue for police and policing functions is the question of arrest and detention authority. At present, all of Liberia's security organizations except the Ministry of Defense have this authority. This is confusing and can lead to abuse. When numerous organizations are authorized to arrest and hold, families and friends of prisoners have difficulty ascertaining the location and status of the prisoner. It is also

---

[5] Because of the specialized training required, the QRPU may be exempt from these rotations. However, this would risk its development as an "elite" force, the sort of force that has had an unfortunate history in Liberia. The possibility of at least some rotations through the QRPU should be considered.

more difficult to ensure that people are apprehended for a legal cause and to ensure that they receive due process—a situation further exacerbated by the problems of Liberia's justice system. Human rights abuses and political arrests are too easy under such a system.

It might be appealing to decide that only the police should have arrest and detention authority. However, if the SSS, Customs, and BIN are not under the LNP, it makes sense that they have the capacity to arrest and briefly hold individuals apprehended in the line of duty. The alternative is to have police officers accompany all other organizations with these functions, which would be inefficient. Limiting the time non-LNP personnel can detain prisoners would help cut down on abuse and ensure appropriate treatment of detained individuals. Thus, one option is to permit these agencies, and these agencies only, to carry out arrests; to limit detention by them to a specified, short period of time; and to consolidate detention authority with the LNP, which would be responsible for jailing prisoners and ensuring that their cases are appropriately dealt with by the justice system.

In all cases, arrest and detention practices must be clearly delineated by law, and police oversight bodies should take on as a primary mission to ensure that these laws are followed and respected. Rules on warrants, evidence, and time limits on detention prior to trial must be established and enforced—and must apply to all organizations with arrest or detention authority. Clear and transparent oversight requires regular reports to the legislature and to the President on practices and events. These reports must also be made easily available to the public.

An argument can be made that intelligence personnel should also have the right to arrest and detain individuals apprehended in the line of duty. This could be especially valuable in the apprehension of high-value targets, such as enemies of the state, about whom intelligence personnel may receive information that no other agency has, when rapid action is necessary, and when coordination is impractical. It can be argued that without arrest authority for intelligence personnel, such targets could easily be lost.

The downside of arrest and detention authority for intelligence agencies is the potential for abuse. Liberia's history of various security agencies being utilized to support personal and political interests is such that there is good reason to believe that an intelligence agency directed by an unscrupulous future government could well abuse these authorities. Moreover, arrest and detention authority for intelligence services, even if carefully circumscribed, blurs what should be a clear line between intelligence and law enforcement. The stringent oversight and transparency requirements of detention described above may be more difficult to implement in the case of intelligence services.

The global record on arrest and detention authority for intelligence functions is mixed. For example, U.S. domestic intelligence agencies have the right to detain; however, Britain's MI-5 does not. The alternative to detention authority for intelligence agencies is a close working relationship with police, including specialized units that include both intelligence and police personnel, whose job is specifically to apprehend targets based on time-sensitive intelligence information.

It is unrealistic for Liberia at this juncture to have the means for interagency mechanisms in every circumstance. Thus, the intelligence personnel could have the authority to arrest individuals who present national-security threats in a clearly defined set of circumstances detailed in law, provided that the arrestee is handed over to the LNP for processing within a strict, fixed time period.

Police Oversight

As democratic policing has evolved over time, most national and local police organizations have determined that three distinct oversight functions are required to ensure effectiveness, professionalism, and accountability. The first oversight function is that of *management*—resourcing, recruitment, equipping, pay, and other support and administrative functions. The second relates to the question of *professionalism*—making sure that the police maintain the appropriate standards in their organization and operations and that those standards are used appropriately by the government. Finally, there is the question of *public accountability*—allowing the public to see that police are there to serve them and to protect their civil liberties, not to violate them.

These oversight functions can be provided for in various ways, with roles for government, the public as a whole, civil society organizations, and institutions within the police. In Liberia's case, several options exist for each area of oversight.

With regard to management, Liberia should consider two possible approaches. First, the Ministry of Justice could take on the responsibility for LNP management. Under this option, the police chief's accountability to the government is strong, and LNP needs are represented in the cabinet. This option precludes direct public input into the definition of police strategy, objectives, and priorities. It also creates some potential for politicization, since the police report through an appointed official.

Another option is the creation of a board to oversee LNP management. This board could be appointed by the President and approved by the legislature and would, by statute, require the representation of members of civil society, specialists, and various communities, as deemed necessary. Under this option, a higher level of public and civil society input is ensured. But this option creates an additional body, which may be cumbersome, complicated, and potentially confusing. In Liberia's difficult post-conflict environment, reliance on a board could result in inadequate government control and attention and a lack of advocacy for LNP in the cabinet. Therefore, this option is inferior to the first one.

In either case—board management or MoJ management—it must be clear that the LNP itself retains operational control of its forces and structures—neither the MoJ nor the board would have authority to deploy, direct, or command units or structures. As with other issues, this must be clearly delineated in statute and the charters for the relevant organizations.

Several options exist for professional oversight and public accountability. First, a single independent board with a national public ombudsman could perform both functions. Deputy ombudsmen in each of the counties of Liberia could liaise with the Monrovia-based national ombudsmen as well as local LNP officers and local boards to increase accountability outside of Monrovia. This is an efficient, simple option that would provide independent oversight of the police.

A second option is to have both functions performed within the LNP. For example, an inspector general could conduct professional oversight as part of that office's regular functions,[6] and an internal national ombudsman could address public accountability in conjunction with county deputy ombudsmen. Although this is an efficient option, it might have less credibility and public confidence because it reduces the independence of police oversight.

A third option is to create an independent board for professional oversight and a separate national ombudsman for public accountability (with county deputy ombudsmen). This option ensures that both functions receive adequate attention and allows for independence and credibility. But it is complicated and potentially cumbersome.

Regardless of the option selected, a few factors must be considered. First, the board would not have management or operational control over the LNP, the former being up to the Justice Ministry and the latter up to the LNP. Second, county-based policing boards should be established to serve as representatives of their communities and to work with the LNP officers assigned to the area in developing policing priorities. The concerns of an overcrowded township in Monrovia, for example, will not be those of a small town in Grand Gedeh. This community-based approach will increase needed community involvement in local policing as well as county-appropriate policing priorities.

Third, the accountability oversight mission must be understood to be one of responsiveness to the public, the legislature, and the state as a whole. Not only must the office of the national ombudsman and the board or inspector general respond to complaints, they also must be able to respond without impediment to requests for investigation by the office of the President, by the legislature, and by members of the public. They need to have the capacity to launch periodic assessments of police departments at the national and local level, and to carry out those assessments on a regular basis, selecting the subjects randomly and ensuring both broad coverage and uncertainty—so that no police

---

[6] An Inspector General's office is required regardless, but in other options it would be responsible only for internal complaints and investigations, with responsibility only to the LNP.

station knows when to expect an investigation. They should also be required to publish regular (e.g., semiannual) reports that address the state of the police today, including the fight against corruption, citizen perceptions, reports of abuses, appropriateness of arrest and detention actions, and government uses of the police. These reports should be provided to all members of the legislature and be publicly disseminated, accompanied by media appearances and fact sheets. Finally, the structures and roles of the office of the national ombudsman and the board or inspector general must be clearly delineated by statute, perhaps through charters for each organization.

The public accountability oversight function in most options does not include hiring and firing authority. It can, however, still be very effective simply by bringing abuses and problems to light, thus enabling the legislature and the executive branch to take action and creating public pressure for them to do so. Liberia should also consider establishing oversight/ombudsman bodies for other security force structures, including the MoD, BIN, and Customs.

Borders

Complete control over Liberia's borders is not feasible. It is therefore crucial to establish priorities and to ensure that the various elements involved in border control complement and collaborate with one another. The current system, in which many organizations and agencies keep representatives at various border crossings—but with little understanding of their respective roles and functions—creates confusion while contributing little to Liberia's security.

The first step in improving border controls is to clearly define the functions of all those with responsibilities at the borders and to publicize these functions broadly, so that Liberians and visitors to Liberia are aware of them. Assuming that police functions are consolidated in the LNP, the agencies involved in border controls could include the BIN, Customs, the AFL, and LNP. To most effectively address the border security threats given resource constraints, the Liberian government should consider putting most Customs and BIN personnel at key crossing points, where most people and goods are likely to cross. These personnel need to have appropriate policing authorities, including arrest and short-term detention until individuals can be transferred to police detention facilities or expelled.

Patrols along the border can be the responsibility of either military forces alone or combined police and military patrols. If police are not involved in patrols, they must be stationed conveniently to ensure that arrest and detention procedures are implemented appropriately. Ground and sporadic air surveillance—the latter provided not by Liberia itself but instead by either donors or contractors—can monitor unofficial border crossings. Liberia's security intelligence service should include as one of its key missions that of monitoring reports of unusual movements at or near borders. This mission should be undertaken in close coordination with the LNP to ensure appropriate responses as well as complete information.

Intelligence

At present, both the MNS and the NSA have intelligence responsibilities. The DEA and NBI have some intelligence responsibilities as well. While most of their missions fit into LNP functions, some aspects of these agencies' work may be more appropriately placed under the auspices of national intelligence. A thorough review of specific functions will be needed to ensure that when and if these organizations are dissolved, their responsibilities and staff are appropriately reallocated or eliminated.

We identify two primary options for restructuring the intelligence sector. First, the MNS could be eliminated, with some of its functions and personnel incorporated into the NSA, as appropriate, along with any relevant components of the DEA and NBI. The NSA would remain Liberia's single national intelligence agency, maintaining its existing staffing and structures. The NSA's mission, which should be clearly defined through a charter for the organization, would be to collect and analyze intelligence, from strategic to tactical, of national security significance. The NSA should not be involved in law enforcement or policy formulation. It would report to the President, advise the security decisionmaking body, and directly support the LNP and the MoD. Given the array of internal security dangers Liberia faces, it is especially crucial that the NSA and the LNP develop and maintain the closest possible relationship from top to bottom.

This option has several advantages. It establishes an independent intelligence service, and—by ensuring that the NSA is an agency, not a ministry—it clarifies NSA's advisory role, ensuring that it cannot be conflated with a policy function. It also enables the Liberian NSC to benefit from common intelligence input, ensuring that the same information and analysis are available to all. The use of the NSA as the sole intelligence agency leverages existing intelligence capabilities. The disadvantages of this option include the potential for operational friction with the police and the functional separation from the police, particularly if both police and the NSA retain arrest or detention functions. This is not insurmountable, but careful management would be required to ensure effective cooperation between the two. A comprehensive public information campaign would be needed to ensure that the Liberian public is well informed about the NSA's role and function as an intelligence agency and about the constraints upon it.

A variant of this option would have the NSA report to the Minister of Justice rather than to the President. This would facilitate a closer link between intelligence and the LNP. It would also concentrate intelligence and armed power in a single ministry. If the NSA has arrest authority, this would be particularly problematic. The NSA would be once removed from the NSC and the President, with the risk that the Ministry of Justice would serve as a filter for intelligence reports.

The NSA should be comparatively modest in size, with perhaps 200 to 300 permanent staff. Its primary intelligence collection focus at the outset would likely be within the country and near its borders. Because the information it would collect would probably be useful for a number of other agencies, coordination and information-sharing will have to be established with the Ministry of Justice, the LNP, the Ministry of Defense, the Ministry of Foreign Affairs, the Ministry of Finance, and others. External intelligence-gathering will be an NSA role but probably would not be considered a high priority at present.

Although the NSA has the primary intelligence function, other agencies would also gather information through the normal course of their work. The police will collect information on crime through daily operations and investigation. They should share with the NSA any intelligence relevant to national security. Regular working groups and liaison should be established between the LNP's branches and the NSA. The Ministry of Defense will be responsible for military tactical intelligence, and it should also have information-sharing mechanisms with the NSA. Because of the importance of intelligence sharing, the NSC should insist that it take place and continually monitor whether it is taking place. The NSA should be responsible for coordination.

Integrated Architecture

Figure 5.1 depicts an integrated architecture for Liberia's security sector. On the left is the national-security decisionmaking apparatus; in the center are the security forces; on the right are the core security functions. In this security architecture:

- The NSC, chaired by the President as commander in chief, has final authority over all security forces.
- Security forces report through ministries rather than directly to the President.
- Security forces are evenly distributed between the Justice and Defense ministries.
- Lines of authority are clear.
- Control over the military passes from the President through the Minister of Defense.
- The number of distinct security forces and services is manageably small, while still allowing for specialization.
- No security force lacks an important core security function.

Figure 5.1
Integrated Architecture and Core Functions

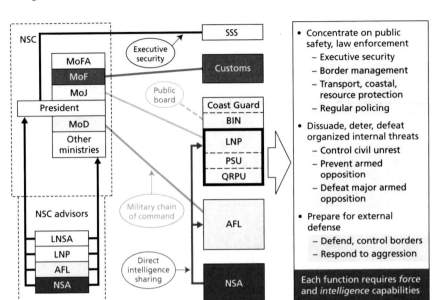

RAND MG529-5.1

- No core security function lacks a force that is principally responsible for it, and there is no confusion or duplication in the alignment of forces with functions.
- The QRPU can support other police units or support the AFL.
- The intelligence service (NSA) reports to the President, services the NSC as a whole, and provides direct support to the LNP and to the AFL.

Although this architecture does not include all of Liberia's current security structure components, it does reorganize those that are the focus of this report. Table 5.2 lists the current organizations and assigns a new configuration for them under the recommended architecture.

Table 5.2
Disposition of Security Organizations

Organization	Proposed Structure
BIN	Separate agency under MoJ
Coast Guard	Created under MoJ
Customs	Separate agency under MoF
DEA	Disbanded; policing functions incorporated into LNP
LNP	Maintains LNP and PSU; develops QRPU
MCP	Disbanded; functions incorporated into LNP
MoD	Oversees AFL and serves on NSC (currently in development)
MNS	Disbanded; functions and appropriate staff incorporated into NSA
NBI	Disbanded; functions incorporated into NSA and LNP as appropriate
NSA	Maintained
NSC	Developed as national security decisionmaking body
SSS	Reports to Office of the President
Additional security forces (FDA, LPRC, LSP, RIA, Telecommunications Corporation)	Disbanded; overlapping functions incorporated into appropriate agencies (LNP, BIN, Customs) Security guard functions: Option 1: Developed into private security forces for relevant enterprises Option 2: Auxiliary police units under LNP Option 3: Part of LNP rotation

CHAPTER SIX

Other Issues

Our analysis of Liberian security forces and institutions has given rise to several issues for further consideration: (1) changes in the legal framework for security; (2) policing priorities; (3) justice, prisons, and the courts; (4) personnel issues; and (5) international security cooperation.

Legal Framework for Security

The legal framework within which the security sector operates is important for a successful transformation. Adoption of a single, new national security law on the authorities, rules, and relationships is one way to approach this. Such a new law would

- create and communicate a clear, consistent framework
- draw in the legislature and facilitate political buy-in
- promote public education, legitimacy, and support
- avoid confusion of multiple potentially contradictory laws and regulations
- facilitate orderly amendment.

An alternative to a new omnibus national security law is to amend existing laws. This may seem simpler in that it enables consideration of issues case-by-case. It also avoids the danger of having to introduce multiple omnibus laws, as they are rejected by Congress on one or another set of grounds, slowing the reform process. Moreover, such an approach would be time-consuming, tedious, and of little use in communicating the breadth, significance, and coherence of individual measures. Moreover, it would run the risk of inconsistency and might fail to gain the public consensus that a comprehensive package could attain. Finally, if a complete law can be agreed to by both the executive and legislative branch, the benefits of legislative acceptance will carry over throughout the reform process.

Another option is to institute the new security sector architecture by presidential decree. Even if this could be done more or less completely within existing law, it would signal a return to executive branch domination and indifference to views of the public representatives.

In the context of codifying the new security sector architecture and authorities, three legal topics deserve particular mention. First, it is important to specify what the specific roles and missions of each agency and office are, including the question of which agencies are to have arrest and detention authority.

Second, the security law should clarify the relationship between the security services and the political process with the aim of avoiding both the interference of the services and their personnel in politics and the politicization of the services. Security forces personnel should have the right to vote. They should not be permitted to campaign for a political party or party seat or hold a political office while serving. Because membership in a political party connotes active partisanship in Liberia, as it does in much of Africa, it is best that security personnel not be party members. Senior security forces personnel should have to wait at least one electoral cycle after retiring before running for office, lest newly retired senior officers be able to exploit their military standing for political gain.

Third, appointments for key roles within the security sector are particularly important. One option is for the assignment of top officials and officers to continue to be conferred through political appointments. This would perpetuate the pervasive politicization of the security sector that has characterized Liberia in the past and has served it so poorly. An alternative is for top security professionals (at least deputy ministers, senior officials, and senior military and police officers) to be nonpolitical appointments, nominated by the President and confirmed by the legislature to serve for fixed terms independent of changes to the government. Although this option may face some political opposition, it does ensure independent security assessments and shields these agencies from politics, even as they answer to elected officials. These officials would include the Chief of Intelligence, the Chief of Staff of the Armed Forces, and the Chief of Police—all three of whom should be advisors to the President and to the NSC. An exception would be the LNSA, which—while also expected to offer objective advice to the President and the NSC—may change with each new government.

The role and authority of the legislature are important for Liberia's national security reform. There are a number of areas in which the legislature plays a significant role. As already discussed, these include decisions with respect to the domestic use of the military and approval of appointments of top advisors, senior officials and officers, and board members. The approval of the legislature should also be required for authorization of the national security budget. In light of past abuses and the need for national consensus on how to provide security in the new Liberia, getting the role of the legislature right is essential for legitimacy and effectiveness.

Any new national security law must take into consideration Liberia's constitution. It does not appear to the authors of this study that constitutional specificity on security sector matters would present a roadblock to the legislative options we have proposed. An omnibus national security law containing all the desired components could be developed, with the legislature enacting those aspects that could be constitutionally enacted by statute. Through an executive order, constitutionally questionable sections can be adopted as a matter of policy, using the broad powers granted to the President by the constitution. As part of an ongoing process of constitutional reform, provision should be made to amend the constitution as needed to allow for eventual legislative enactment of those security sector reforms adopted by executive order. However, a team of Liberian constitutional specialists, perhaps assisted by counterparts from abroad, should examine these issues as part of developing the national security legal framework.

Policing Priorities

At present, the LNP have few resources. Bicycles, motorbikes, flashlights, radios, and batons are few in number and poorly distributed. The LNP is unarmed and has had limited training. Consequently, the LNP is most effective when it is consistently and visibly accompanied by UNPOL when conducting patrols in Monrovia and other major towns. In light of the current limitations of the LNP, UNPOL may need to obtain arrest authority to provide proper support. This would necessitate a UN Security Council decision. Even without arrest authority, a greater UNPOL presence on patrols would bolster the effectiveness of the LNP, simply by ensuring that an armed foreign officer is standing behind LNP personnel when they confront criminals.

The LNP would be much more effective if they had adequate logistics—particularly means of transportation and communication. Radios and bicycles or motorbikes would be particularly helpful, as would flashlights to support nighttime patrolling.

In rural areas, the focus of the LNP and UNPOL should be on high-priority and high-risk locations. In the longer term, greater mobility is key to improved rural policing. Resources permitting, LNP officers should conduct regular patrols along the roads to facilitate community policing. Although many Liberians live great distances even from county seats where the LNP is based, most live within a reasonable distance of one of the roads. If it were known that at a regularly scheduled interval a community relations patrol would be moving slowly along the main road, the citizens in rural areas could avail themselves of this occasion to interact with the officers and, quite possibly, pass along useful information. Police patrols in those areas should not, however, be limited to scheduled visits; they should also be carried out at random and, to the extent possible, respond to incidents when called. Police should generally make themselves available to local citizens in the course of their duties.

In the near term, private security will probably need to protect large operations such as rubber plantations and mines. Certainly a mechanism for accrediting and monitoring private security is important. Private security providers should not have arrest and detention authority; they need to work with local police.

Justice and Courts

Effective policing requires functional and legitimate courts and prisons. The legitimacy of the security sector depends on the justice system. Without an efficient, fair, and transparent system of arrest, trial, and incarceration, those accused of crimes may simply be jailed without due process or, at the other extreme, find their way right back onto the street without prosecution.

Donor assistance to jump-start and sustain justice reform is essential. The International Crisis Group recently completed recommendations for short-, medium- and long-term reforms in the justice sector that deserve significant consideration.[1] The RAND team has identified two particular areas of concern: the prisons and the courts. In the near term, temporary holding facilities need to be built. In the long term, Liberia should build integrated police, detention, and court facilities in each of the fifteen county seats. There is an immediate need for foreign advisors (ideally, Liberian expatriates with the appropriate expertise could be identified) to assist Liberian judges, as well as for an exploration of donor funding of salaries and other reforms. Oversight of the judiciary is needed. For the long term, Liberia needs to create a sustainable system for handling court cases, including filing, providing books of statutes to judges, setting up and operating court calendars, and making transcriptions of proceedings. While some broad institutional reform efforts in this sector are now under way, this area has remained woefully under-resourced. Unless this situation changes, the problems and pathologies it engenders run the risk of undermining a great deal of the broader security sector reform effort.

---

[1] ICG (2006c).

Redundancy

Throughout Liberia's security sector, there are entirely too many individuals drawing paychecks but not contributing appropriate effort. Agencies are overmanned; some have numerous "ghost workers" on the payroll—fictitious names that are used by employees to gain two or three additional salaries or employees who only report to work to collect their paycheck. Overstaffing hinders the creation of a coherent, effective, affordable national security sector. As recent experience with the LNP indicates, reform and training will be rendered meaningless unless redundant personnel are retired.

The current government's focus on "right-sizing" is highly important for the future of the security sector and more broadly, for the government. Although funding has now been secured to provide severance packages for redundant LNP staff, similar redundancy programs are needed for other ministries and agencies, which must go through the same sort of stringent vetting, recruiting, and training. As the security sector is streamlined and rationalized, the individuals who work in existing agencies will need to have a straightforward, coherent process through which they can either (1) apply for work at successor or replacement agencies, (2) be retired, or (3) be helped to find new employment. Current security sector employment must not be a guarantee of future employment in this sector, and, as organizations are combined, staff must not be moved en masse. Rather, each individual must be separately considered for vetting and training in the reformed security sector. This applies to new structures, such as the Coast Guard, and continuing ones, such as BIN and Customs.

Wages

The system of paying wages needs to be fixed. Currently, government wages are paid sporadically. Government employees outside of Monrovia have to travel to the capital to collect their paychecks. Paychecks sometimes "go missing," and there are reports of police officers who demand a portion of subordinates' checks before they release them.

The Ministry of Finance needs to make payroll its top priority. The new wage disbursement system needs to be transparent and automatic. Once the armed forces are established, a quartermaster, rather than individual officers, is to be in charge of distributing wages to soldiers. A similar system should be instituted for the police. Publicly paying all government employees at the same time helps prevent the paymaster from demanding bribes for providing a paycheck or from expropriating part of a worker's wages. Once the banking system becomes solvent, automatic deposits will be the preferred method of payment because this method reduces the risk of corruption, prevents withholding of paychecks, and in general improves security. Branches of Liberia's central bank or other commercial banks can be used to pay wages outside of Monrovia.

International Security Cooperation

The goal of Liberian security sector transformation is for the state to be able to create and maintain safe and peaceful conditions for the people while also contributing to a more secure environment in the immediate region. But it is neither feasible nor advisable for Liberia to achieve this goal without international partnership. For some years to come, international cooperation will be essential for security sector transformation and for security itself. Even for the longer term, Liberia needs allies and partners—and it should cultivate them now.

The Mano River Union, established to promote economic cooperation and integration in the region, should be reinvigorated and built upon, eventually promoting security cooperation, such as through joint border controls and other confidence-building measures. Liberia should also continue to cooperate closely with Sierra Leone, as both countries continue on the path to general and security sector reform. They will have opportunities both to learn from and to support each other's efforts. Regional cooperation in such areas as oceans and waterways, surveillance, and border controls can yield significant cost savings, efficiencies, and security improvements for all.

Politically and practically, multilateral security cooperation is on the rise in Africa as a whole and in West Africa especially. Liberia would do well to involve itself in these efforts. Liberia should begin to play an active role in ECOWAS, including participating in activities and even peace-keeping operations, on a very modest scale, when it has the capacity to do so. By the same token, Liberian membership in the African Union can enable it to expand its network of ties and partnerships beyond the subregion. ECOWAS and the AU also provide appropriate multilateral settings for Liberian cooperation with other African states, e.g., Nigeria.

The United Nations will have a role in Liberia for some time. It is crucial that UNMIL maintain a presence until Liberia has developed proven security forces and institutions—and possibly, to a smaller degree, even beyond that point. Throughout UNMIL's tenure and beyond, the UN Security Council should continue to monitor developments in Liberia, making it clear to potential enemies that the Council's interest in Liberian security will not recede with the gradual downsizing of UNMIL. The passage of new resolutions at important junctures would confirm that the Security Council is monitoring progress and remains committed.

A formal alliance relationship or binding defense commitment from the United States is unlikely. But the United States will remain a friend to Liberia and Liberia should leverage that friendship. It will benefit not just from U.S. aid, but from security cooperation that is visible to neighbors and others and that can assist Liberia with specific needs, such as intelligence support. This relationship can be built both through continuing the efforts already under way and through a program of port calls, military assistance, and training, which should be a priority for Monrovia as it defines its relationship with Washington. Finally, as Liberia and its friends seek additional partners and assistance, their efforts should conform to the principles, architectures, and standards of Liberia's emergent security sector.

CHAPTER SEVEN

Key Findings and Implementation Priorities

Forces

- The existing Liberian–UN–U.S. plan to build a small LNP and small AFL provides a necessary and useful baseline, but it may prove inadequate to satisfy Liberia's security needs, especially to ensure basic public safety and prevent armed internal opposition.
- Increasing the size of the LNP, while also introducing a quick-response police unit (QRPU) of the LNP and a small Coast Guard, would better meet Liberia's security needs and would reduce its dependence on domestic intervention by the AFL. This option would increase annual operating costs above the current plan by about $6 million. The capital cost of this option could be roughly $35 million more than that of the current plan. This option seems like a very good investment for Liberia and its partners—returning greater security at an affordable operating cost.
- While the planned AFL is small by regional standards, it is neither necessary nor advisable to decide now how much if at all larger it should be. For now, the emphasis should be on quality.
- The ability of Liberian security forces to combat armed gangs and insurgency will be hampered by poor roads, especially during rainy seasons. This problem can be reduced by good surveillance, rotary-air mobility (provided by a foreign partner), preemptive action, and the isolation of armed groups in inaccessible areas. Improving Liberia's roads is important for its security.

- It will take at least five years before these Liberian forces are fully built, equipped, trained, and deployed. During that period, it should be possible to scale back significantly the numbers, and thus the cost, of UNMIL provided certain core UNMIL capabilities are preserved— especially police advisors, the UNMIL quick-response force, airlift, and air surveillance.
- During the lengthy transition from UNMIL to Liberian security forces, respective command and control systems must be compatible and must enable coordinated or combined operations.
- In the longer term—perhaps ten years or so—there will remain a small but critical need for international military capabilities to complement Liberian forces, especially advisors and rotary-wing lift and surveillance.
- Liberian force plans must be fully resourced, continually reviewed in light of the evolving security environment, and adjusted as necessary. The uncertainty associated with this environment places a premium on adaptability.
- In time, Liberia must develop its own ability to assess its needs for forces and other capabilities based upon informed, objective, and realistic planning. Creating a civil-military ability to assess and align resources with needs should be a part of the assistance Liberia receives from its international partners.

Organizing Government

- A Liberian NSC is needed for policymaking, resource allocation, and crisis management. It should be chaired by the President and should include at its core the Ministers of Justice, Defense, Finance, and Foreign Affairs. It should receive professional advice and objective analysis from the head of national intelligence, the most senior officers of the LNP and AFL, and from the Liberian National Security Advisor).
- The NSC system should be extended downward from the cabinet level to working levels to ensure interministerial cooperation. The LNSA and staff should guide this coordination.

- The chain of command over the AFL should flow from the President, as commander in chief, through the Minister of Defense, to the top general, with the understanding that decisions to use military force should be reached by NSC deliberation and, for domestic use of the AFL, in consultation with the legislature.
- The Liberian government should consolidate ancillary police into the LNP, with the exception of a small Special Security Service (under the office of the President), the Bureau of Immigration and Naturalization, and a Coast Guard (under the Ministry of Justice), and Customs (under the Ministry of Finance).
- Within the LNP, qualified police should be rotated to the extent practical among the regular police to the PSU and the QRPU.
- The LNP should come under the authority and management oversight of the Ministry of Justice. It should also have an independent board to ensure professional excellence, apolitical conduct, and public trust. It will retain operational control of its forces.
- Intelligence capabilities and activities must be held to the same standards as all other security sector activities. Responsibility for collecting national security intelligence should be concentrated in the National Security Agency, which should report to the President, provide analysis to the NSC, and directly support the LNP and AFL. This intelligence service should have a tightly restricted authority to arrest and briefly detain persons who pose a national security threat. Recognizing that the police will be able to collect much of the information needed to fight crime, the NSA should be of modest size and focus on high-level concerns.
- Such sweeping changes should be codified in an unambiguous manner that delineates missions and roles and that secures broad political buy-in, earns public understanding and acceptance, and avoids future misunderstandings. An omnibus national security law is the best way to meet these needs.
- Liberia should continue to seek assistance with the broad package of security sector reform. In doing so, it must ensure that all assistance is aligned with its overall reform plan and the standards and requests it establishes.

Special Issues

- Apart from voting, security personnel should stay out of politics.
- Senior officials and officers should be nominated by the President and confirmed by the legislature.
- As they are being developed and rebuilt, the LNP need to be accompanied on patrols by armed international police advisors.
- Liberian justice and courts systems must be built quickly or law enforcement will be neither effective nor legitimate.
- Personnel of the old police force and other existing security forces who will not be trained and integrated into the new forces should be retired, so they do not infect the new forces with old, bad practices.
- Current systems for paying security personnel must be upgraded and made immune to corruption.
- Liberia must not be and need not be left to struggle with its security challenges alone. Even as Liberian forces gradually replace UNMIL, and as sound security institutions are built, those with a stake in Liberia—the UN, the AU, ECOWAS, the United States, other countries and international organizations, and even Liberia's Mano River Basin neighbors—can and should help.

Immediate Implementation Priorities

- The NSC should begin functioning regularly and without delay.
- In addition to its regular duties, the NSC should have cognizance over the implementation of security sector transformation plans.
- The U.S. government, UN, and Liberian government should begin consultations on amending current force-building plans, with particular attention to
  - creation of a QRPU as a branch of the LNP
  - creation of a Coast Guard
  - enlargement of the planned LNP.
- The U.S. government, the UN, and the Liberian government should begin developing a common multiyear plan that encom-

passes both the building and fielding of Liberia's security forces and the level and capabilities of UNMIL.

- Steps should be taken to ensure the presence of an UNMIL quick-response force until Liberia develops such a capability.
- The new design for security forces and services should be discussed in the NSC and with other well-known Liberian figures and with the general public, with particular attention to deconfliction and cooperation and to the alignment of forces and services with ministries.
- Legal and substantive experts from Liberia and its partners should be engaged to frame a new national security law under the direction of the NSC.
- Plans should be created promptly for standing up courts, appointing and ensuring oversight of judges, establishing prison facilities, retiring redundant workers, and ensuring the payment of wages.

Capacity Building

It is hoped that this work will enable Liberia, with the help of the United States, the UN, and other stakeholders, to fashion an integrated architecture for a transformed security sector. Following the logic of this analysis, this effort should begin with a set of governing principles and criteria; include an assessment of Liberia's security environment and its concept for addressing the challenges of that environment; identify core security functions, forces to fulfill those functions, and institutions to manage those forces; and present the new state's ideas to the public for discussion and eventual codification.

This is a demanding agenda, one that cannot be tackled by a handful of officials and staff of the new government, however much international help they receive. To succeed both in implementing Liberian plans and in providing security within a new system, it is essential to greatly expand the number of officials, officers, politicians, and other leaders who understand the principles that underlie it and the practices that animate it.

To this end, it is strongly recommended that Liberia and the U.S. Department of Defense develop, without delay, specific plans to educate ministers, officials, officers, and others—for example, through the Africa Center for Strategic Studies at Fort McNair, Washington,[1] complemented by in-country courses. This education should be based on existing courses as well as on activities specifically geared toward meeting Liberia's own particular challenges.

The transformations suggested in this report may face powerful opposition from those with vested interests in the status quo. Rational change, as is often the case, has no institutionalized constituency precisely because the institutions have not yet been built. Security sector transformation requires stewardship of the process, steering it through the hurdles that lie ahead, assuring the integration of various components into the overall architecture as they come in, and keeping the whole enterprise on course. While President Johnson Sirleaf must be the steward, she will need great and steady help from home and abroad.

---

[1]   Other organizations, such as the U.S. Institute of Peace, may also have relevant programs.

West African Military Balance

Table A.1
Armed Forces per Capita in West Africa

Country	Population	Armed Forces	Citizens per Soldier
Benin	7,649,360	4,550	1,681
Burkina Faso	13,491,736	10,800	1,249
Cape Verde	418,224	1,200	349
Côte d'Ivoire	17,298,040	17,050	1,015
The Gambia	1,595,086	800	1,994
Ghana	21,946,247	7,000	3,135
Guinea	9,452,670	9,700	975
Guinea-Bissau	1,413,446	9,250	153
Liberia (small army)	2,900,000	2,042	1,420
Liberia (large army)	2,900,000	4,748	611
Mali	11,415,261	7,350	1,553
Mauritania	N/A	N/A	N/A
Niger	12,162,856	5,300	2,295
Nigeria	128,756,768	78,500	1,640
Senegal	11,706,498	13,620	860
Sierra Leone	5,846,426	12,500	468
Togo	5,399,991	8,550	632
Average	17,753,758	13,298	1,285

SOURCES: IISS (2006) and CIA (2007).

Table A.2
Armed Forces per Square Mile of Territory in West Africa

Country	Area (mi^2)	Armed Forces	Mi^2 per Soldier
Benin	43,483	4,550	9.56
Burkina Faso	105,869	10,800	9.80
Cape Verde	1,557	1,200	1.30
Côte d'Ivoire	124,503	17,050	7.30
The Gambia	4,363	800	5.45
Ghana	92,456	7,000	13.21
Guinea	94,926	9,700	9.79
Guinea-Bissau	13,946	9,250	1.51
Liberia (small army)	43,000	2,042	21.06
Liberia (large army)	43,000	4,748	9.06
Mali	478,767	7,350	65.14
Mauritania	397,955	N/A	N/A
Niger	489,191	5,300	92.30
Nigeria	356,669	78,500	4.54
Senegal	75,749	13,620	5.56
Sierra Leone	27,699	12,500	2.22
Togo	21,925	8,550	2.56
Average	137,936	13,298	10.37

SOURCE: IISS (2006).

Costing

RAND estimated both operating costs and capital costs for the different options presented in this report. Operating costs were split between wage and nonwage support costs. Wage costs were calculated based on the current pay of $90 per month for enlisted men and women and for police officers.

Nonwage support costs were based on detailed cost components provided in the spreadsheets and annexes supporting the study by MPRI for the U.S. Department of Defense (MPRI, 2004). The per-capita nonwage support costs were calculated from the detailed data on supplies, fuel usage, and other items for each type of force. Per-capita figures were multiplied by force size to estimate nonwage support costs.

Capital costs were calculated from the same source. The structure, major equipment, and size of the force were used to estimate costs for military units. Per-capita cost estimates were used in conjunction with force sizes to estimate the capital costs of police units.

Bibliography

Alao, Abiodun, John MacKinlay, and Funmi Olonisakin. *Peacekeepers, Politicians, and Warlords: The Liberian Peace Process*. Foundations of Peace, United Nations University Press, 2000.

Bekoe, Dorina A. "Toward a Theory of Peace Agreement Implementation: The Case of Liberia." In Rose M. Kadende-Kaiser and Paul J. Kaiser, eds., *Phases of Conflict in Africa*, Willowdale, Ontario: de Sitter Publications, 2005.

Campbell, Greg. *Blood Diamonds*. Boulder, Colo.: Westview Press, 2004.

Central Intelligence Agency. *The World Factbook 2007*, Washington, D.C.: Central Intelligence Agency, 2007.

CIA—*See* Central Intelligence Agency.

Conteh, Al-Hassan Conteh, Joseph S. Guannu, Hall Badio, and Klaneh W. Bruce. "Liberia." In Adebayo Adedji, ed., *Comprehending and Mastering African Conflicts: The Search for Sustainable Peace and Good Governance*, New York: Zed Books in association with African Centre for Development and Strategic Studies, 1999.

Davies, Victor A.B. "Liberia and Sierra Leone: Interwoven Civil Wars." In Augustin Kwasi Fosu and Paul Collier, eds., *Post-Conflict Economies in Africa*, New York: Palgrave Macmillan, 2004.

Dunn, Elwood D. "Liberia's Internal Responses to ECOMOG's Interventionist Efforts." In Karl Magyar and Earl Contein-Morgan, eds., *Peacekeeping in Africa: ECOMOG in Liberia*. New York: Palgrave Macmillan, 1998.

Ellis, Stephen. *The Mask of Anarchy: The Destruction of Liberia and the Religious Dimension of an African Civil War*. New York: New York University Press, 2001.

———. "How to Rebuild Africa." *Foreign Affairs*, Vol. 84, No. 5, 2005, pp. 135–148.

Ero, Comfort. "Dilemmas of Accommodation and Reconstruction: Liberia." In Michael Pugh, ed., *Regeneration of War-Torn Societies*, New York: Palgrave Macmillan Ltd., 2000.

Global Witness. "An Architecture of Instability: How the Critical Link Between Natural Resources and Conflict Remains Unbroken." December 2005. As of February 10, 2006:
http://www.globalwitness.org/media_library_detail.php/144/en/an_architecture_of_instability

Human Rights Watch. "Youth, Poverty and Blood: The Lethal Legacy of West Africa's Regional Warriors." Human Rights Watch Report, Vol. 17, No. 5, March 2005. As of February 10, 2006:
http://hrw.org/reports/2005/westafrica0405/

ICG—*See* International Crisis Group.

IISS—*See* International Institute for Strategic Studies.

International Crisis Group. "Conflict History: Liberia." September 22, 2004. As of April 15, 2006:
http://www.crisisgroup.org/home/index.cfm?action=conflict_search&l=1&t=1&c_country=64

———. "Liberia's Elections: Necessary but Not Sufficient." Africa Report No. 98, September 7, 2005. As of February 10, 2006:
http://www.crisisgroup.org/home/index.cfm?id=3646&l=1

———. "Liberia: Staying Focused." Africa Briefing No. 36, January 13, 2006a. As of February 10, 2006:
http://www.crisisgroup.org/home/index.cfm?l=1&id=3872

———. "Guinea in Transition." Africa Briefing Number 37, April 11, 2006b. As of April 15, 2006:
http://www.crisisgroup.org/home/index.cfm?id=4067&l=1

———. "Liberia: Resurrecting the Justice System." Africa Report Number 107, April 6, 2006c. As of April 15, 2006:
http://www.crisisgroup.org/library/documents/africa/west_africa/107_liberia_resurrecting_the_justice_system.pdf

International Institute of Strategic Studies. *The Military Balance, 2006,* London: Oxford University Press, 2006.

MPRI—*See* Military Professional Resources, Inc.

Military Professional Resources, Inc. *Sustainment Budget: New Armed Forces of Liberia,* Alexandria, Va.: MPRI, 2004.

Pham, J. Peter. Liberia: *Portrait of a Failed State.* New York: Reed Press, 2004.

———. "U.S. National Interests and Africa's Strategic Significance." *American Foreign Policy Interests,* Vol. 26, 2005.

———. "Hesitant Home Repair or Successful Restoration? Foreign Policymaking in the George W. Bush Administration, the Conflict in Liberia, and the Case for Humanitarian Non-Intervention." In Glenn P. Hastedt and Anthony J.

for Humanitarian Non-Intervention." In Glenn P. Hastedt and Anthony J. Esterowicz, eds., *The President and Foreign Policy: Chief Architect or General Contractor?* New York: Nova Science, 2005.

———. "Reinventing Liberia: Civil Society, Governance, and a Nation's Post-War Recovery." *The International Journal of Not-For-Profit Law,* Vol. 8, No. 2, January 2006. As of April 15, 2006:
http://www.icnl.org/knowledge/ijnl/vol8iss2/art_2.htm

Prkic, Francois. "The Phoenix State: War Economy and State Formation in Liberia." In Klaus Schlichte, ed., *The Dynamics of States: The Formation of Crises of State Domination,* Aldershot, England, and Burlington, Vt.: Ashgate, 2005.

Ruohomaki, Olli. "The Bad Governance Challenge Facing Fragile States." In *Development in an Insecure World: New Threats to Human Security and Their Implications for Development Policy,* Helsinki: Global.Finland, 2005. As of February 10, 2006:
http://global.finland.fi/julkaisut/pdf/DevelopmentSecure.pdf

Sawyer, Amos. *Beyond Plunder: Toward Democratic Governance in Liberia.* Boulder, Colo.: Lynne Rienner, 2005.

Save the Children. "Fighting Back: Child and Community-Led Strategies to Avoid Children's Recruitment into Armed Forces and Groups in West Africa." As of February 10, 2006:
http://www.savethechildren.org.uk/scuk_cache/scuk/cache/cmsattach/3487_Fighting_Back.pdf

Sesay, Max. "Security and State-Society Crises in Sierra Leone and Liberia." In Caroline Thomas and Peter Wilkin, eds., *Globalization, Human Security and African Experience,* Boulder, Colorado: Lynne Rienner, 1999, pp. 145–161.

United Nations. "Financing of the United Nations Mission in Liberia—Performance Report on the Budget of the United Nations Mission in Liberia for the Period from 1 August 2003 to 30 June 2004." Report of the Secretary-General, No. A/59/624, December 20, 2004.

U.S. Department of State, Bureau of African Affairs. "Background Note: Côte d'Ivoire." June 2006a. As of August 10, 2006:
http://www.state.gov/r/pa/ei/bgn/

———. "Background Note: Guinea." May 2006b. As of August 10, 2006:
http://www.state.gov/r/pa/ei/bgn/

———. "Background Note: Senegal," April 2006c. As of August 10, 2006:
http://www.state.gov/r/pa/ei/bgn/2862.htm

Twaddell, William. "Testimony by William H. Twaddell, Acting Assistant Secretary of State for African Affairs, Hearing on Liberia Before the House International Relations Committee." June 26, 1996. As of February 10, 2006:
http://dosfan.lib.uic.edu/ERC/bureaus/afr/960626Twaddell.html

Yoder, John Charles. *Popular Political Culture, Civil Society and State Crisis in Liberia*. Lewiston, N.Y.: E. Mellen Press, 2003.